FLYING SCOTSMAN

FLYING SCOTSMAN

THE MOST FAMOUS STEAM LOCOMOTIVE IN THE WORLD

JAMES S. BALDWIN

First published 2013

The History Press
The Mill, Brimscombe Port
Stroud, Gloucestershire, GL5 2QG
www.thehistorypress.co.uk

British Library Cataloguing in Publication Data.
A catalogue record for this book is available from the British Library.

ISBN 978 0 7524 9447 0

Typesetting and origination by The History Press
Printed in India

Front cover image: No 4472 *Flying Scotsman*, complete with appropriate headboard, is seen working the 'Cumbrian Mountain Express' over the Settle & Carlisle Railway during the early 1980s. (David Ward collection)

Back cover image: No 4472 *Flying Scotsman* at one of the many stops on its North American tour. During its 'First Mission in North America', 1969, the exhibition train travelled through seventeen states of eastern and southern USA. (Paul Dowie collection)

I dedicate this book to my mum and dad, who awakened in me the desire to look closely and observe what is all around us, which many do not notice.

On 27 May 2004, *Flying Scotsman* and its support coach worked a positioning run from Southall to Doncaster, on the first leg of its transfer to the National Railway Museum at York. The author is seen alongside *Flying Scotsman* at 05.40 at Southall, just prior to its departure from its home depot for the very last time in private ownership. (Colin Crisp)

CONTENTS

ACKNOWLEDGEMENTS

Quite some time ago I decided that a complete reference work of the steam locomotive *Flying Scotsman* had to be compiled. I set about putting together the vast jigsaw of its known history into a single book, and began interviewing as many people as possible who have been associated with this locomotive, before it was too late – as none of us is getting any younger!

To this end, I wish to thank the many people associated with *Flying Scotsman* who have answered my questions, provided images or items of memorabilia for me to examine and who have recalled personal stories of this magnificent machine.

In particular, I wish especially to thank the owners of *Flying Scotsman* through the preservation years:

Alan Pegler OBE; Sir William McAlpine, 6th Baronet; Dr Pete Waterman OBE; Dr Tony Marchington; and finally, Andrew Scott CBE and Steve Davies CBE, both former directors of the National Railway Museum (NRM). They have all been amazingly helpful in allowing me unbridled access to their private collections and archives, and in giving me the time to question them for minutiae of their memories and memorabilia. Without them, this book would not have been possible.

Here are some of the many people that have contributed personally to my quest to complete this amazing account, without whom some of the best bits about *Flying Scotsman*'s history would have been lost for ever:

Helen Ashby OBE; Chloe, Harriett and Steff Baldwin; Alan Battson; Tom Blake; Ian R. Bolton; Kyle Bosworth; Henry W. Brueckman: Peter Butler; Terry Bye; John Cameron; Robert Carroll; Tom Carter; David Chapell; Christopher Chesney; David Court; Geoff Courtney; Colin Crisp; Maria Christodoulou; Derek Crunkhorn; Carole Cuneo; Petrina Derrington; Dennis Dickens; Steve Doughty; Paul Dowie; Nev Evans; Kevin Fisher; Antony Ford; Pat Fortescue; Edward Gardner; Tony Gooding; Peter Grafton; Ian Gunn; Peter Hall; Ray Harris; Dr Ian Harrison; George and Richard Hinchcliffe; Chris Hogg; Alex Ingram; David Jones; Michael Kemp; Roland Kennington; Ivor and Mike Laws; Mat Lenihan; Dave Maffei; Philip Mason; Stuart Matthews; Bob Mitchell; Khaled Monib; Patrick Mullee; John Neville; John Newman; Martin K. O'Toole; Penny Pegler; Grahame Plater; Cara Randall; Terry Robinson; Margaret Ritchie; David Rollins; D. Trevor Rowe; Edward Saalig; Harry Scotting; Tony Shooter; Roger Sinar; Brian Southon; Bob Stadelman; Bob Stanyard; Dave and Fred Stenle; Jim Styles; Ted Talbot; Peter N. Townend; Mike Watts; Joy and Richard Woods; and David Ward.

Thank you each and every one of you.

I also thank my wife Harriett for her unfailing encouragement.

This book about *Flying Scotsman*'s history has set out to document the evolution of this magnificent locomotive and to bask in its glorious achievements. It has the amazing ability to bring out romance, nostalgia and emotions in all of us. May *Flying Scotsman* continue to bring enjoyment, happiness and an occasional tear to one's eye during its future adventures.

Flying Scotsman – The People's Engine. I'll raise a cheer to that!

James S. Baldwin
London, 2013

INTRODUCTION

I first became aware of *Flying Scotsman* during my formative years, when I was a choirboy in Pimlico, London. One day, during a break in the choir's rehearsal for the following Sunday, one of my fellow choristers presented me with a copy of an Ian Allan loco-spotting book. My home was near to Grosvenor Bank, just out from Victoria Station, so I didn't need much persuasion in underlining the numbers of passing trains.

I remember being awakened several times from my slumbers at some witching hour by my dad, who wanted me to witness the passing of many a train being assisted up the bank by struggling tank engines, leaving volcanic plumes of steam and smoke in the night air!

As well as steam engines, I noted down the numbers of ordinary trains arriving and departing from Victoria and soon became entranced with some of the more exotic names, such as the 'Night Ferry' and the 'Brighton Belle'. I particularly remember seeing the luxurious Pullman cars of the 'Golden Arrow' train being worked by *Britannia* and Bulleid 'Pacifics'. Just prior to its arrival at Victoria Station, I noticed the smartly attired Pullman car attendants polishing the grab rails of the Pullman cars, as the train drifted down the bank into Platform 8.

All of the above is probably why a name like *Flying Scotsman* would have caught my attention, so imagine my surprise and delight when it was announced that the most famous steam locomotive in the world was going to depart from my very own local station, Victoria, on 17 September 1966.

The day in question arrived and the group of train-spotters in my area were in a state of great excitement as No 4472 *Flying Scotsman* backed down Grosvenor Bank to couple up with its train. I was truly mesmerised by this highly polished, bright green, giant steam locomotive, complete with its brightly painted red nameplates, as it dominated the scene. After a wait that seemed to go on forever, watching the all-too-short departure of this incredible machine had me hooked for life.

A few days later, when I was watching the BBC children's programme *Blue Peter*, I learned that their presenter John Noakes had been on the footplate of No 4472 on that very trip to Brighton and had indeed been filming the story of the last steam-hauled non-stop run from Victoria to Brighton – for me this was indeed a very special day to remember.

On a different occasion, I remember that I 'skipped off' school to witness another non-stop run by *Flying Scotsman*. This time, on 1 May 1968, No 4472, together with its second tender, departed at 10.00 from King's Cross with the 'Fortieth Anniversary Non-Stop Special', bound for Edinburgh Waverley. As this legendary locomotive departed, again watched by an amazing number of people – young and old alike – this was another event for me to remember forever. Little did I realise that *Flying Scotsman* would subsequently prove to become such a major part of my life, with me making films and writing books about it to such a great extent.

In a previous era, the 'Flying Scotsman' train, which had been named after the locomotive, had included in its formation such luxuries as a cocktail bar serving, what else, but the 'Flying Scotsman Cocktail', which was invented in the 1920s by Harry Craddock in the legendary American Bar at the Savoy Hotel in London.

So why not make yourself your very own 'Flying Scotsman Cocktail' to enjoy while you read the rest of this book? You will need: a chilled martini glass, 25ml of light Scotch whisky, 20ml of Midori, 10ml of dry sherry and 25ml of green apple juice. Shake the contents, pour into the chilled martini glass and decorate with a cape gooseberry.

So now suitably imbibed, we return to the 'Flying Scotsman' train, which also boasted a cinema and, to make you look smart, a hairdressing salon, although the thought of having a short back and sides or even a close shave as the train was thundering over point work would surely make one's hair stand on end!

It seems that everyone has heard of *Flying Scotsman*, but what exactly do they mean by the name? To answer this question correctly we need to look at two parts.

First, there is the 4-6-2 steam locomotive that was designed by Nigel Gresley and which was the third member of the A1 class to be constructed at Doncaster Works. It was numbered 1472, cost £7,944 to construct and was the first express passenger locomotive to be completed for the newly formed London & North Eastern Railway (originally L&NER, but almost immediately becoming LNER) in 1923.

On 27 December 1923, because of a fractured centre piston rod, No 1472 entered Doncaster Works for repairs, while there was no immediate replacement. Also, as it wasn't going anywhere for some time, it was decided to prepare the engine as the LNER's exhibit for the British Empire Exhibition at Wembley. No 1472 had by now been allocated the number 4472 and painted in the new LNER house livery of apple green. It had an LNER coat of arms mounted on each side of the locomotive, brass trim was added to the wheel splashers, the locomotive's tyres and motion were highly burnished and all copper and brass fittings were brightly polished.

No one quite knows when the name *Flying Scotsman* was first coined or by whom, but it was at this time that No 4472 was fitted with *Flying Scotsman* nameplates.

For display at the Wembley Exhibition, No 4472 was coupled to its own tender, which had the letters LNER surmounting the number 4472 on the sides. After preparation was completed, and in order to maintain its pristine condition for its journey to Wembley, it was wrapped up in a full-length hessian cover to maintain the bright paintwork.

Over the following decades, *Flying Scotsman* would make many more public performances, cementing its place in the public awareness as 'the most famous steam locomotive in the world'.

An un-rebuilt Battle of Britain class 'Pacific' locomotive prepares to depart from Victoria with a down 'Golden Arrow' service to Folkestone in the early 1960s as a small boy – probably about the same age as the author – looks on. Scenes such as this inspired the author to become interested and involved with steam railways. (Peter N. Townend)

To answer the second part of the question – 'What is the *Flying Scotsman*?' – we turn to the 'Flying Scotsman' named rail service that still runs even today on the East Coast Main Line between London and Edinburgh. The service consists of passenger-carrying carriages, working on an express service between King's Cross and Edinburgh Waverley. The train was actually named after the *Flying Scotsman* locomotive and not vice versa, as is sometimes thought. This prestigious and long-standing King's Cross to Edinburgh Waverley express had been popular since June 1862 and was known then as the 'Special Scotch Express'. It had simultaneous departures at 10.00 from the Great Northern Railway's (GNR) King's Cross and the North British Railway's Waverley Station.

The 10.00 departure, which was renamed as the 'Flying Scotsman', first appeared in the LNER's public timetables on 11 July 1927, along with the 'Aberdonian' and 'Night Scotsman' services. On 1 May 1928, No 4472 *Flying Scotsman* became the first locomotive to actually work the regular 'Flying Scotsman' service, which initially ran non-stop between London and Edinburgh. In 1930, rolling stock used on this service included a 'Louis XVI-style' restaurant, while one carriage enabled passengers to listen to the radio through headsets and see radio-photographs of such events as the finish of the Derby.

So we see there are in fact two Flying Scotsmen, but they are connected!

• • •

Flying Scotsman is often referred to as a 'Pacific' class locomotive, but what does this actually mean?

Well, to call a steam locomotive a 'Pacific' is to use a sort of railway shorthand to describe the wheel arrangement of this particular type of steam locomotive, as was originally used to describe a system devised by Frederick Methvan Whyte. It first came into use in the early twentieth century and was encouraged by an editorial in the *American Engineer and Railroad Journal* of December 1900.

Whyte's system first counts the number of leading wheels, then the number of driving wheels and finally the number of trailing wheels of a locomotive, with the groups of numbers being separated by dashes. So in the 'Whyte Notation', as it is known, a locomotive with two sets of leading axles – hence four wheels in front – then three sets of driving axles (with six wheels) and finally one trailing axle (two wheels) is classified as a '4-6-2'. The 4-6-2 type of wheel arrangement was first used on locomotives that had been built by the Baldwin Locomotive Company of America for use in New Zealand – and New Zealand's coastal Pacific location led to the 'Pacific' name being adopted for a 4-6-2.

• • •

Number 4472 *Flying Scotsman* has many notable accomplishments, which will be covered in the rest of this book. Amongst other achievements, it was the locomotive chosen by the LNER to represent the best that they had achieved in locomotive design at the Wembley Exhibition of 1924 and 1925; it was the engine which worked the first regular non-stop run of the 10.00 'Flying Scotsman' express in 1928; and was the 'star vehicle' in the UK's first foray into the world of 'talkie' feature films.

In 1934, No 4472 *Flying Scotsman* managed to top all of these by recording the first officially authenticated run of 100mph under steam power, when it worked a test train down Stoke Bank. This record has been improved upon over the years, with many praiseworthy exploits by other locomotives. The London, Midland & Scottish Railway (LMS) claimed the British railway speed record for steam traction when they announced that their 'Pacific' No 6220 *Coronation* had reached a maximum speed of 114mph on the descent of Madeley Bank, while hauling a special press train on 29 June 1937, with Driver T.J. Clarke and Fireman C. Lewis on the footplate. However, this claim was based on a reading of the Hasler on-board train monitoring system in the cab

› Part of an advert advertising the cocktail bar on the newly launched 'Flying Scotsman' train, which left King's Cross and Edinburgh Waverley simultaneously at 10.00 during the late 1920s. (Author's collection)

^ No 4472 *Flying Scotsman* is seen with the train of the same name during the late 1930s. (W.B. Greenfield, courtesy of the NELPG)

and could not be substantiated, as there was no official 'paper trail' to back up the reading. As this high speed run was coming to an end No 6220 had insufficient braking distance remaining before entering a series of crossover points at Crewe, putting it in severe danger of derailing. Although the train held the rails, much crockery in the dining car was smashed. This incident caused the LMS and LNER to agree to stop these dangerous record-breaking runs, which had come to be seen by the public for the publicity stunts that they were.

The *Coronation*'s acclaimed run of 1937 was soon eclipsed by LNER's A4 class 'Pacific' locomotive *Mallard*, which was a streamlined development of the *Flying Scotsman* type of locomotive designed for better efficiency at high speeds. It officially reached an amazing 126mph on a run in 1938, on the downhill section of track at Stoke Bank, claiming the world speed record for steam traction that it holds to this day.

Despite all of this, it was somehow the locomotive *Flying Scotsman* which captured and retained fame in the public imagination, even though it was replaced by more modern forms of motive power on express services. Despite losing its historic number, 4472, to become first No 502, then No 103, then E103 and, later still, British Railways (BR) No 60103, as well as losing its magnificent apple-green livery for 'war-time' black, and then drab BR corporate green, the magic survived.

When the locomotive was scheduled for withdrawal by BR in 1963, Alan Pegler, who was a former member of the British Railways Board Eastern Region and a businessman in his own right, sold the family business, bought *Flying Scotsman* and, as a private individual, ran the locomotive all over the country. He repainted *Flying Scotsman* in its well-known and admired LNER apple-green livery and restored its original number – No 4472. Alan Pegler, in saving *Flying Scotsman*, won the hearts and gratitude of many people of all ages and different walks of life, and saved what many consider to be the best of the age of steam on Britain's railways for future generations.

For a short period, *Flying Scotsman* was the only one of the many preserved steam locomotives that was officially allowed to run on the British Railways network, since the official abolition of steam traction on British Railways in 1968. However, there was a lot of bad feeling amongst other locomotive owners, who were not allowed to run their engines on BR tracks. So, in winter 1968/9, No 4472 *Flying Scotsman* was prepared for a promotional tour of North America during 1969. As will be seen in chapter 6, *Flying Scotsman* was prepared extensively for running on North American railway lines, and shipped across the Atlantic on board *Saxonia*.

On its arrival, *Flying Scotsman* worked a 'British Trade Exhibition Tour Train' around the USA and Canada. Although the tour was initially successful and garnered a lot of publicity, financial difficulties led to *Flying Scotsman* ending up on the frontage at Fisherman's Wharf in San Francisco as a static exhibit.

^ Numbered 60103 and wearing corporate dark-green livery, *Flying Scotsman* passes through Hitchin, working a King's Cross to Leeds service during 1957. (Frank Hornby)

› The 'power and the glory' of No 4472 *Flying Scotsman* is seen to good effect as it prepares to depart from Newcastle Station in the mid-1930s. (W.B. Greenfield, courtesy of the NELPG)

In October 1972 an article from the *Lubbock Avalanche Journal* detailed the woes that running the locomotive caused Alan Pegler:

> Most people love a steam locomotive, but few really want one of their own. Alan Pegler, 52, is the exception. He is in love with the Flying Scotsman, the first railroad train ever to surpass the 100 mile per hour mark – or as the British say, 'to do a ton'. Pegler's love affair with the legendary immigrant from England has cost him $1 million and now he may not be able to afford her. The Flying Scotsman awaits her fate on a siding, her green and red paint baking in the sun, tourists uninterested. The engine failed as a tourist attraction in San Francisco when Pegler was forced to move from a choice location at Fisherman's Wharf. The result has been bankruptcy for the English businessman who brought his train to America. Negotiations are under way to move the train to Southern California for inclusion in a subdivision around the *Queen Mary* liner, converted into a convention centre and tourist attraction at Long Beach. But first Pegler must plead his case in the London bankruptcy court … Now No 4472 is the lone survivor of 79 engines of her type. She sits cold and silent on a rail siding at Sharpe Army Depot five miles south of Stockton.

The move from San Francisco to Stockton for safe storage in a secure military depot was a last ride 'wake', with some forty friends and railway enthusiasts aboard the train for the 40-mile trip to Lathrop. Participants took turns blowing *Flying Scotsman*'s whistle as they played at engineer in the plundering cab and scrambled through the tender corridor, which had originally enabled the engine to run 393 miles non-stop between the English and Scottish capitals in its heyday. They sat reflectively in *Lydia*, a Pullman car used by Churchill and Eisenhower during the Second World War, or just joined the party under way in the observation car.

Eventually *Flying Scotsman* was transferred, with the accompanying carriages, to Sharpe Army Depot in California for storage away from the eyes of the public. Located in a military base during the Vietnam War, this perhaps explains why so few details, photographic or otherwise, have emerged of the locomotive's time there.

Pullman car *Lydia* was sent back to its rightful owners at the Railroad Museum at Green Bay, Wisconsin, mounted on a 'low-boy' rail wagon and, together with sister Pullman car *Isle of Thanet*, has since been repatriated.

During the period after storage of the train commenced at Sharpe Army Depot, Phil Monte, a local railway scrap dealer, acquired two of the exhibition train carriages (namely Exhibition Car A and Exhibition Car B). They had, like the rest of the train, been seized by the Internal Revenue Service (IRS) in lieu of unpaid debts that had been accrued by the 'Flying Scotsman' on tour, which were owed to a local Californian railway operator. Phil attempted to sell the coaches as garden storage sheds while they still contained boxes of *Flying Scotsman* literature. The bodywork of these coaches was in poor condition, with severe leaking apparent from the roofs. Unfortunately, when no one showed any interest in them, Phil stripped the cars of their bogies and equipment, burned the wooden bodies and sold the under frames to a Californian farmer, who used them as bridges over a creek on his farm!

The five exhibition vehicles left over remained stored at Sharpe Army Depot until they were bought by the Victoria Station Restaurant Company for further use. These vehicles were then sent to a shopfitter's workshop in California, where they were refurbished and modified. At the workshop, three of the five carriages were converted into dining cars by having

^ On 13 November 1965, No 4472 *Flying Scotsman* worked the 'Panda Pullman' fund-raising special return from Paddington to Cardiff. Including a 7-minute water stop at Swindon, the special arrived into Cardiff at 16.59, having broken the record for the fastest run of a steam locomotive from Paddington to Cardiff. No 4472 also broke the return record on its way back to London. (D. Trevor Rowe)

› A young Bill McAlpine stands in front of his new acquisition. (Sir William McAlpine collection)

windows cut into the bodywork. The remaining carriages were for use at the new Victoria Station Restaurant located at Universal Studios, California.

The observation car initially went to Universal Studios as well, but was later moved to the headquarters of the Victoria Station Restaurant Company in San Francisco, where it was used as the boardroom at their Chestnut Street offices. It has since been repatriated and now resides at the Swanage Railway.

The remaining four coaches stayed in use as dining cars at Universal Studios from 1977 until 1997, when the restaurant received a major makeover and the four coaches quietly disappeared. Despite making several enquiries with staff at the complex, including speaking to the vice president of Universal Studios and taking many tours of the vast back lots in this amazing studio complex, all traces of the four coaches have gone. I have also made extensive enquiries with many preserved and operational railway establishments in California and the surrounding states, but I have found no information as to the subsequent whereabouts or fate of these four coaches. It would appear, therefore, that they have all been scrapped.

In December 1972, Bill McAlpine was approached by Alan Bloom, the steam enthusiast and developer of Bressingham Gardens, to see if he would mount an attempt to rescue *Flying Scotsman* from America. Bill realised that a personal assessment of the situation was vital and so George Hinchcliffe, who had been operational manager for *Flying Scotsman* on the North American tour, was asked to go to the United States and see what was required to get No 4472 back to England.

So after a lot of quick organisation, by 8 January 1973 a bill of sale for 'Flying Scotsman' had been drawn up, in which the locomotive, its two tenders and associated spares were to be purchased by Bill McAlpine for $72,500. A ship was found and No 4472 was moved from Sharpe Army Depot to a naval supply depot at Oakland.

During the Second World War, the depot was a major source of supplies and war materials for ships operating in the Pacific. The depot had its origin in 1940 when the Navy bought 500 acres of wetlands from the city of Oakland for $1. The Navy reclaimed the land and populated it with large warehouses and opened it on 15 December 1941, rapidly expanding the site over the decades. In the late 1940s it was renamed Naval Supply Center, Oakland, and the name later changed again to Fleet and Industrial Supply Center, Oakland. During the Cold War, it was one of the Navy's most important supply facilities and was closed in 1998.

So No 4472 was moved to the dockside at Oakland and was prepared in readiness for its lifting from American soil. After No 4472 *Flying Scotsman* and its two tenders were loaded on board the *California Star*, they were literally welded to the ship's deck to ensure that they were not lost at sea during the long journey home via the Panama Canal.

In a press release dated 22 January 1973, it was announced that:

> Class A3 4-6-2, No 4472 *Flying Scotsman* and its two tenders, formerly owned by Mr A.F. Pegler and stored at a US Army Depot in Stockton, California, was loaded on board the ship the *California* Star at Oakland and was expected to arrive in Liverpool in early February. It had been purchased by Mr W.H. McAlpine. After unloading, the locomotive was to go to a British Rail Engineering Limited workshop for inspection and repainting. Subsequently, it will be put into store prior to display in the National Railway Museum in York – before the museum was due to open, from where it is intended to use it to power steam-hauled excursion trains.

The total distance run by No 4472 *Flying Scotsman* in North America was approximately 15,400 miles. The estimated cost of purchasing No 4472

During the summer of 1993, *Flying Scotsman* left Babcock Robey Works after a heavy repair, having been repainted in its former BR green livery. With the running number changed to No 60103, it then worked on preserved railways between 1993 and 1996. (D. Trevor Rowe)

A full-size drawing of a bronze axle box, made at BR's Doncaster drawing office, as used by engineers at Riley & Son during repair work to *Flying Scotsman* in 2012. Notably, the class is called 'A1' and the year of manufacture is 1949. (Author)

During the process of overhauling *Flying Scotsman* it was painted in 'war-time' black livery during testing. Here is the side of the tender showing the letters N and E, when it was receiving attention during its 2012 repair. (Author)

to the satisfaction of its new owner and returning it to this country was believed to be in the region of £60,000.

Flying Scotsman is the only steam locomotive to have circumnavigated the globe, having travelled through the Panama Canal on its return from California and then in 1988, it travelled via the Cape of Good Hope to Australia and broke yet another world record, by working a train non-stop for 442 miles.

After its travels to North America and Australia, *Flying Scotsman* went back to the main line, working excursions around the British network. This proved to be the beginning of many more adventures for this iconic locomotive.

Flying Scotsman then received an overhaul around this time, during which a double chimney and smoke deflectors were refitted and it reverted to its former BR livery of dark green, also taking its former number, 60103.

It then spent several years running on preserved railways, until it was withdrawn from service during 1995.

In 1993, pop music producer Pete Waterman bought, from Sir William McAlpine, 50 per cent of 'Flying Scotsman Enterprises', which had previously run all the steam-hauled specials on the old BR network, thereby giving him a half share of the world's most famous steam locomotive. Shortly thereafter, along with Sir William, Pete, as CEO of the new company, purchased the first BR company to be sold off in privatisation, which was the 'Special Train Unit'.

In 1996, *Flying Scotsman* was bought by Dr Tony Marchington, who paid for *Flying Scotsman* to have the most expensive and extensive overhaul ever carried out to a steam locomotive – in private ownership. After the overhaul was completed, *Flying Scotsman* put up some of the best performance runs of its career, particularly when working heavy Venice-Simplon Orient Express (VSOE) Pullman trains and when Tony sold *Flying Scotsman* in 2004, it was in a much better condition than when he purchased it.

And even after all this, there was still more to come!

In 2004, No 4472 *Flying Scotsman* was sold to the NRM in York where, after a short period working 'Scarborough Spa Express' excursions, it entered into a very prolonged overhaul, which cost even more than the previous overhaul carried out during Tony Marchington's ownership!

In May 2010, *Flying Scotsman* once again caused much public interest and debate by being painted in 'war-time' black livery, with the letters NE on the tender.

^ George Stephenson's original *Rocket* survives in a much-altered state in the Science Museum in London, but this is the Baltimore & Ohio Railroad replica as displayed in Chicago in 1893. (Author's collection)

● ● ●

I said earlier that *Flying Scotsman* was 'the most famous steam locomotive in the world', but does it really deserve to wear this title? Is it more famous than *Rocket*, *Mallard* or even *Thomas the Tank Engine*? Well if it's not, then it jolly well should be; with so many superlatives associated with this locomotive, over such a wide time span, there are so many good reasons for *Flying Scotsman* to hold this title. They certainly don't make them like that anymore – although there is *Tornado*, of course, but that is a different story and it is after all a derivative of *Flying Scotsman*.

Through the actions of people like Alan Pegler, Sir William McAlpine, Pete Waterman and Tony Marchington, *Flying Scotsman* has been saved for the nation, for future generations to admire and enjoy.

But whatever happens to *Flying Scotsman*'s well-being and appearance, the passion in people is still stirred all over the world by the sight and sound of this locomotive. However, due to the many overhauls and component changes carried out during its colourful life, there is very little, if anything, remaining of what was on that first locomotive, which entered service with the L&NER back in 1923. For example, *Flying Scotsman* has received seventeen different boiler changes in its long and illustrious life and has worked with nine different tenders.

What is left, however, is the name *Flying Scotsman* and the thoughts this magical name conjures up. It was *Flying Scotsman* that was always featured in children's picture books of previous generations; the crack, high-speed express locomotive of the East Coast railway route between London's King's Cross and Edinburgh Waverley via York. It was *Flying Scotsman* that took some of us to Scotland for our summer holidays. The name conjures up the spirit and excitement of what was best about British engineering after the First World War, when Britain led the world in innovative design and where technology was pushed to its very limits.

Flying Scotsman is a real and tangible link with our glorious past and should be an inspiration to all in the future.

Note:

Throughout the book, to avoid confusion where locomotives and trains are known by similar names, locomotive names are given in italic type (e.g. No 4472 *Flying Scotsman*) while named trains, services and classes of engine are given in single quotes (e.g. 'Flying Scotsman', 'Pacific' class).

1

WHY WAS *FLYING SCOTSMAN* BUILT?

The 'Flying Scotsman' is the present name of a daily British express passenger train service that has been running between London and Edinburgh, the capitals of England and Scotland, since 1862. In 2013 this service was still running. The East Coast Main Line, over which the 'Flying Scotsman' service runs, was built in the nineteenth century by many small railway companies. But mergers and acquisitions led to three main companies controlling the route: the North British Railway (NBR); the North Eastern Railway (NER); and the Great Northern Railway (GNR).

In 1860, these three companies established the East Coast Joint Stock, a group of common vehicles for use on through services, and it was from this agreement that the 'Flying Scotsman' service came about. East Coast Joint Stock provided rolling stock that was jointly built and maintained by the three major companies listed earlier. The L&NER, which absorbed both the GNR and NER in 1923, kept many of the old 'joint stock' carriages in use until nationalisation in 1948.

The first 'Special Scotch Express', as it was known in 1862, took 10½ hours to complete the journey and included a half-hour stop at York for lunch. This disadvantaged the poorer passengers, who suffered greatly as they tried in vain to purchase and consume their scalding hot soup in an almost comical rush.

Then, over two summers in the late nineteenth century, British passenger trains belonging to different railway companies literally raced each other from London to Scotland over the two principal rail trunk routes, the East Coast and the West Coast, with the rivalry for supremacy becoming known as the 'Race to the North'. As rail travel became more popular, competition for custom increased even further and so, along with the not inconsiderable improvements in technology, the overall time from London to Edinburgh was reduced to 8½ hours by 1888, an improvement of 2 hours.

From 1900, the King's Cross to Edinburgh services were improved once more, with the actual trains being made more customer friendly. New features incorporated in the trains that later became standard were corridor connections between the carriages and the unheard of luxury of heating inside the passenger compartments. The introduction of dining cars included in the GNR trains was great news for passengers, as they could now take unhurried luncheon on the train if they wanted to. The bad news was that even though the York stop had been reduced to 15 minutes, there was even more of a feeding frenzy for passengers as they tried to consume their hot food in the limited time allowed, and to add insult to injury, even accounting for a shorter stop, the through end-to-end journey time still stayed at 8½ hours.

› A view looking into the cab of a GNR 'Atlantic' locomotive. The 'Atlantic' class was the predecessor to H.N. Gresley's 'Pacifics'. (Author's collection)

›› This 1922 timetable shows services between Glasgow/Edinburgh and London. 1922 was the year that the first of H.N. Gresley's 'Pacific' locomotives was completed and entered service. (Author's collection)

GLASGOW, FALKIRK, POLMONT, LINLITHGOW, and EDINBURGH.—North British.

Week Days.

Miles	Up.	mrn	mrn	mrn	mrn	mrn	mrn	mrn	mrn	mrn	mrn	mrn	aft	aft	mrn	aft
			Restaurant Car Express. Glasgow to Edinburgh and Leith.	Arrives Grangemouth at 9 25 mrn., see page 779.	Thro' Restaurant Cars, (commencing 10th inst), Glasgow to Southampton, see pages 778, 735, 731, 669, 698, 82, 47, and 128. Through Carriage, Glasgow to London (K.C.)	Express, Glasgow to Edinburgh and Leith.	Commences on the 10th instant.	Not after the 8th instant.	Arrives Grangemouth at 12 32 aft., see page 779.	Through Carriage, Glasgow to London (K.C.), Scarboro' and Penzance, see pages 778, 735, 732, 758, 669, 698, 82, 15, and 22. Restaurant Car, Glasgow to York. Commences on the 10th instant	Not after 8th instant. Through Carriage Glasgow to London (K.C.)	Arrives North Berwick at 2 12 and Gullane at 2 14 aft., see page 778. Sats. only. Lothian Coast Express.	On 15th, 22nd, and 29th instant.	Except the 15th, 22nd, & 29th inst.	Restaurant Car Express.	
	797 Helensburgh ..dep.	....	6 21	....	7 30	9 0	10 0	10 0	....	1047	10e57	11 2			1140	1215
—	Glasgow (Queen St. †).dep.	6 20	7 42	8 20	8 40	10 5	1115	1120	11 45	12 0	12 10	1225			1 0	2 0
1¼	Cowlairs	6 27	....	....		....	1123	1128	....							....
3¾	Bishopbriggs	6 32	....	8 31		....	1129	1134	....							....
6¼	Lenzie	6 39	7 57	8 37		10B19	1136	1141	....							....
11¼	Croy	6 48	....	8 46		....	1145	1150	....							....
12¾	Dullatur §	6 53	....	8 50		....	1150	1155	....							....
15¼	Castlecary §	7 0	....	8 56		....	1157	12 2	....							....
18¼	Bonnybridge	7 7	....	....		....	12 4	12 9	....							....
21¾	Falkirk (High) ‖	7 16	8 18	9H17		....	1212	1217	12H24			1256				2 33
25	Polmont 779, 791	7 25	8 26		9 17	10 44	1218	1223		1237	12 47		1242	1251		2 41
27¼	Manuel 791, 811	7 36	....			....							1248	1258		....
29¾	Linlithgow	7 43	....			10 54							1 d9	1 7		2 50
32¾	Philpstoun	7 50	....			11 v1							1 16	1 14		....
35¼	Winchburgh	7 57	....			11 v8							1 23	1 22		....
39	Ratho 808	8 4	....			....							1 30	1 30		....
41¾	Gogar	8 10	....			....							1 36	1 36		....
45¾	Haymarket 782, 793	8 19	8 54		9 45	11t18							1 45	1 45	2 0	3 13
47¼	Edinbro'* 778, 781.arr.	8 23	8 58		9 49	11t22				1 5	1 15	1 28	1 49	1 49	2 4	3 17
49¾	Leith (Central) .... "	8 36	9 13		1016	11N35				1 22	....	....	2 3	2 9	2 38	3 45
171¾	778 Newcastle (Cn.) arr.	....	....		1256	....				3 54	4 10	....	....	....	6 45	....
252¼	778 York .... "	....	....		2 27	....				5 53	6 4	....	....	....	9 16	....
440¾	778 London (K.C.) ... "	....	....		6 30	....				10 0	10 0	....	....	....	....	....

Week Days.—*Continued.* / **Sundays**

Up.	aft	aft	aft	aft	aft	aft	aft	aft	aft	X	aft	Sun. mrn	Sun. aft
	Arrives North Berwick at 5 32 and Gullane at 5 34 aft., see page 778.	Except Saturdays. Lothian Coast Express.	Restaurant Car Express, Glasgow to Leeds.		Arrives Grangemouth at 5 27 aft., see page 779.	Restaurant Car, Glasgow to Edinburgh.		Restaurant Car Express.		Sleeping Car and Thro' Car-riage, Glasgow to Lond'n (K.C.)		Express, Through Gl'gow to London (K.C.)	Sleeping Car and Through Carriage, Glasgow to London (Kings Cross)
797 Helensburgh ..dep.	1e10	2 0	2845	....	....	3 43	..	4 28	6U42	8 5	....	....	....
Glasgow (Queen St. †) dep.	2 50	3 50	4 0	....	4 28	4 55	5 25	6 0	7 55	9 30	940	9 55	8 55
Cowlairs	2 58			....	4 35		5 32	....	....		946		....
Bishopbriggs	3 4			....	....		5 37	....	....		950		....
Lenzie	3 12			....	4 44		5 44	....	8 10		....		....
Croy	3 21			....	4 53		5 53	....	8 19		....		....
Dullatur §	3 26			....	4 57		5 58	6 22	8 24		....		....
Castlecary §	3 33			....	5 3		6 5	....	8 31		....		....
Bonnybridge	3 40			....	....		6 12	....	8 38		....		....
Falkirk (High) ‖	3 47			....	5H20	5 28	6 20	6 35	8 45		....	1026	9 26
Polmont 779, 791	3 55			....		5 36	6 28	6 42	8 53	h	....	....	....
Manuel 791, 811	4 3			....			6 35	....	9 0		....	....	....
Linlithgow	4 8			5 2			7K12	6 51	9 6		....	1040	9 40
Philipstoun	....			5 9			7 19	....	....		....	....	....
Winchburgh	....			516			7 26	....	9 16		....	....	....
Ratho 808	....			525			7 34	....	....		....	....	....
Gogar	....			533			7 40	....	....		....	....	....
Haymarket 782, 793	....		5 0	542		6 5	7 49	....	9 32		....	....	....
Edinbro'* 778, 781.arr.	....	4 50	5 4	546		6 9	7 53	7 15	9 36	1030	....	11 3	10 3
Leith (Central) .... "	....	....	5 18	625		6 25	8 13	7 49	....	....	....	....	....
778 Newcastle (Cn.) arr.	....	....	7 54	....		....	....	1055	1 X8	1 24	....	1 55	1 8
778 York .... "	....	....	9 42	....		....	....	1 10	3 X1	3 17	....	3 51	3 1
778 London (K.C.) .. "	....	....	....	....		....	....	6 0	7X30	7 35	....	8 15	7 30

* Waverley Station.
† High Level.
§ Station for Cumbernauld (1¾ miles).
‖ Nearly 1 mile to Falkirk (Grahamston) Station, Caledonian.

For other Trains

BETWEEN	PAGE
Falkirk and Polmont	779
Glasgow & Edinburgh 795,	838
Glasgow & Lenzie 788, 793,	805
Glasgow and Dullatur 788,	793
Glasgow and Castlecary	788
Glasgow and Bonnybridge	805
Glasgow and Linlithgow	793
Falkirk and Edinburgh	787
Ratho & Edinburgh 789, 793,	805

*** **For Steamer Sailings** from Watering Places to and from the Clyde, see pages 907 to 911

B Takes up for Edinburgh and beyond.
d Arrives Linlithgow at 12 54 aft.
e Except Saturdays.
h Stops at Polmont when required to take up for London (King's Cross).
H Grahamston Station.
K Arrives Linlithgow at 6 40 aft.
N Arrives Leith (C.) 11 48 mrn. on Weds.
s Saturdays only.
t 6 minutes later on Wednesdays.
U Leaves at 6 50 aft. on Saturdays.
v Wednesdays only.
X Except on the 1st and 8th instant, Sleeping Car, from Fort William to London (King's Cross).
X Except Saturdays, commencing 10th instant. Arrives Newcastle 12 46, York 2 44, and London (K.C.) 7 10 mrn.

MANUEL and BO'NESS.—North British.

Week Days only.

Miles frm Manuel.	Up.	mrn	mrn	mrn	mrn	aft	mrn	non	aft (Except Sats.)	mrn (On the 1st & 8th inst.)	aft	aft	aft		aft	aft	aft	aft	aft		
	790 Edinburgh (W.) dep.	7 40	....	....	10 5	a	1130	a	....	1130	2 0	2 20	3 31	....	....	5 5	5 28	....	8 0	....	....
	791 Glasgow ¶ 793.. "	6 20	7 42	8 47	10 5	....	d	12 0	12F10	....	2 0	....	2 50	....	4 20	4 55	....	6 0	7 55	....	....
	Polmont ....dep.	....	9 3	9 34	1050	....	....	1249	1257	1 5	2 47	....	....	....	5 10	5 44	....	7 8	9 15	....	....
—	Manuel .... "	8 25	....	....	....	1212	1223	....	....	....	....	3 42	4 22	....	....	....	6 13	....	....	....	....
3¼	Kinniel	8 34	9 17	9 48	11 4	1221	1232	1 1	1 11	1 19	3 1	3 51	4 31	....	5 24	5 58	6 22	7 22	9 29	....	....
4¼	Bo'ness ....arr.	8 37	9 20	9 51	11 7	1224	1235	1 4	1 14	1 22	3 4	3 54	4 34	....	5 27	6 1	6 25	7 25	9 32	....	....

Week Days only.

Miles	Down.	mn	mrn		mrn	mrn	a	d	a	d	aft (Sats. only.)	aft (Except Sats.)	aft	aft	aft		aft	aft	aft	aft	
	Bo'ness ....dep.	715	7 50	....	8 45	10 2	12 0	12 2	1230	1242	1 55	2 5	3 28	4 8	4 40	....	5 35	6 37	6 48	8 19	....
1	Kinniel	719	7 54	....	8 49	10 6	12 4	12 6	1234	1246	1 59	2 9	3 32	4 12	4 44	....	5 39	6 41	6 52	8 23	....
4¼	Manuel 790, 811...arr.	728	....	....	....	....	....	....	1243	1255	....	....	3 41	4 21	....	....	....	....	7 1	....	....
6¾	Polmont 790 .. "	....	8 8	....	9 3	10 20	1218	1220	....	....	2 13	2 23	....	....	4 58	....	5 53	6 55	....	8 37	....
31¾	790 Glasgow ¶ 793 .arr.	....	9 0	....	1052	11 15	1 16	1 16	....	....	3 10	3 10	....	....	6 3	....	....	....	....	9 40	....
24	791 Edinburgh*.... "	823	8 58	....	9 49	11A22	1 15	1 15	....	1 49	3 17	3 17	5B46	....	6 9	....	6 50	....	....	9 36	....

a Runs on the 15th, 22nd, and 29th.
A Arrives Edinburgh at 11 28 mrn. on Wednesdays.
B Passengers can arrive Princes Street Station, Edinburgh at 4 43 aft.
d Except on 15th, 22nd, and 29th.
F Leaves at 12 noon, on and after 10th instant.
* Waverley Station.
¶ Queen Street, High Level Station.

Due to a long-standing agreement between the competing East and West Coast main-line railways since the famous 'Railway Races' of 1888 and 1895, the high speeds attained on the so-called 'Scotch Expresses' were limited and the travelling time for the 392 miles between the capitals had been reduced to only a relatively pedestrian 8¼ hours.

With the ever growing pressure of improved timings and more carrying capacity on the 'Scotch Expresses', the GNR was seeking something to lead the market, so they introduced the first 4-4-2 or 'Atlantic'-type locomotive in Great Britain. Designed by Henry Ivatt in 1897, they were built at Doncaster Works and the class were commonly known as 'Klondikes', having been created in the year of the great Klondike gold rush.

Next, larger boilered 'Atlantics' were introduced, but when H.N. Gresley succeeded Henry Ivatt in 1911, no further 'Atlantics' were produced. However, Gresley continued to make improvements to the 'Atlantics', and the fact that the 'Atlantic' design was capable of being bettered in so many ways delayed the need for the GNR to invest in brand-new express passenger locomotives. But with the popularity of rail transport increasing, Gresley eventually came under pressure to do something special.

So the need for a more powerful and faster design led H.N. Gresley to look around for new ideas and his attention was drawn 'across the pond', where he became inspired by the Pennsylvania Railroad's K4s steam locomotive – a true American classic. Proudly nicknamed the 'Standard Passenger Locomotive of the World' by the Pennsylvania Railroad, the K4s class first ran in 1914 and soon became a familiar sight all along the railway as the premiere steam passenger locomotive. No fewer than 425 examples were built by the time production ended in 1928 and the K4s class carried on working right to the very end of 'Pennsy' steam in 1957, having hauled some of the railway's most prestigious passenger services and also freight trains.

With inspiration drawn from the K4s class and many of his own designs, H.N. Gresley produced the first of two of his three-cylinder passenger express 'Pacific' locomotives and put them to work on the GNR in 1922. The first example was numbered 1470 and was subsequently named *Great Northern*. Satisfaction for these new locomotives came early, as an order was placed for a further ten locomotives to be constructed. The third member of the 1470 class was completed at a cost of £7,944 and was delivered in 1923, the year in which the railways of Britain were grouped into the so-called 'Big Four'. It was the first express passenger locomotive to be officially delivered to the newly formed L&NER and, before long, No 1472 was renumbered to became No 4472 and was painted in the new company's house colours of stunning apple-green livery, lined out in black and white.

The 1470 class of locomotive incorporated many new designs and ideas, one of which was the use of 'conjugated valve gear'. Designed by H.N. Gresley and assisted by Harold Holcroft, this valve gear enabled a three-cylinder locomotive to operate with only two sets of Walschaerts valve gear for the outside cylinders and, by using this outside motion via the '2 to 1' lever, it operated the valve motion for the inside cylinder to operate.

No 4472 entered service to great plaudits from inside and outside the rail industry, but due to a mechanical fault it was taken out of service and, as the replacement component was deemed to be a long time in the making, the opportunity was taken to prepare it to exhibition standards. After that, it was given the name *Flying Scotsman*, wrapped up in a bespoke hessian cover and was dispatched from Doncaster to Wembley.

No 4472 *Flying Scotsman* was then put on display at the 1924 British Empire Exhibition at Wembley, providing great publicity for the LNER. Other locomotives on display at the exhibition were: Great Western Railway (GWR) 'Castle' class locomotive No 4073 *Caerphilly Castle* and Stockton & Darlington Railway 0-4-0 *Locomotion* No 1, the first locomotive to run on a passenger railway.

During the months following the Wembley Exhibition, H.N. Gresley announced at the annual dinner of his premium apprentices and pupils that the LNER and GWR would run comparative exchange trials between their two types of locomotives.

In 1925, No 4472 *Flying Scotsman* was put on display to the public at the British Empire Exhibition, Wembley, for a second time. Adjoining the main stands was a group of old and new locomotives, including GWR's

^ A dining car as used in the 'Anglo-Scottish Express' services of the 1890s. The off-centre aisle gave a 'four plus two' table arrangement that was quite common during this time. (Author's collection)

^ With the arrival of the Ivatt 'Atlantic' locomotives after 1898, the Stirling 'Single' class locomotives began to be displaced from the most prestigious express services. Several examples were rebuilt by H.A. Ivatt after 1898 with a domed boiler, but withdrawals of the 1870 series began in 1899. The last examples of the class were in use on secondary services until 1916. The first of the class, No 1, is the only one to have been preserved and is exhibited at the NRM in York. (Author's collection)

^ In 1924 No 4472 *Flying Scotsman* was put on display to the public in the 'Palace of Engineering', Wembley. No 4472 had been coupled to its original tender No 5223, with the letters LNER surmounting the No 4472 on the tender sides. Other locomotives on display at the exhibition included the Stockton & Darlington Railway's *Locomotion* No 1, which is seen on the left. (Sir William McAlpine collection)

^ No 4472 *Flying Scotsman* is coupled to corridor tender No 5323, which replaced its original tender No 5223. The detail of the tender is seen to good effect as *Flying Scotsman* rests on the King's Cross turntable. (Author's collection)

'Castle' class No 4079 *Pendennis Castle*. Again, as in 1924, the Great Western engine was declared to be more powerful than its bigger LNER rival and in the months after the second Wembley Exhibition, the LNER and GWR companies ran further comparative exchange trials.

No 4074 *Caldicot Castle* took part in the GWR–LNER locomotive trials of 1925 against the H.N. Gresley-designed A1 class 'Pacific' No 4474 *Victor Wild*, running between Paddington and Plymouth. During April and May of the same year, GWR No 4079 *Pendennis Castle* also ran trials against H.N. Gresley's A1 class 'Pacific' No 4475 *Flying Fox*, running between King's Cross and Doncaster. However, on its very first run No 4475 failed with a 'hot box' (an overheated axle bearing) and had to be replaced by No 2545 *Diamond Jubilee*. The 'Castle' made smart work of the ascent from King's Cross to Finsbury Park and regularly ran this section in less than 6 minutes, a feat that Gresley's 'Pacifics' were unable to reproduce. The result of these tests was that the 'Castle' class locomotive was shown to be more economical in both coal and water on the test runs, as compared to the A1 class. Its superiority was attributed to a higher boiler pressure and, in particular, better valve gear. The LNER learned valuable lessons from the trials, which resulted in a series of modifications carried out from 1926 on No 4477 *Gay Crusader* and which were then adopted for all of the Gresley 'Pacifics' to improve their performance.

Forty more A1 class 'Pacifics' were built by the LNER between 1924–25. No 4472 returned to normal main-line service in November 1925 and settled down to the regular work of a Doncaster-based 'Pacific' locomotive, working services mainly between Doncaster and London.

So as a direct result of the Wembley exhibition rivalry and the information gained by the LNER from the trials against the GWR, the modifications made to the LNER A1 class locomotive's valve gear caused its coal consumption to be drastically reduced, and it was subsequently found to be possible to run the 'Flying Scotsman' service non-stop with a heavy train on a single tender full of coal.

For the introduction of the non-stop 'Flying Scotsman' service on 1 May 1928, ten specially designed corridor tenders were constructed with a coal capacity of 8 tons. They provided access to the locomotive's footplate from the train through a narrow passageway inside the tender tank. The passageway, which ran along the right-hand side of the tender, measured 5ft high by 18in wide. A flexible bellows connection linking it with the leading coach was incorporated into these tenders. The idea was that it '... prevented engine crew fatigue, by enabling a replacement driver and fireman to take over halfway from London to Edinburgh – but without having to stop the train and so saving time' – or so the publicity

No 4472 *Flying Scotsman* prepares to depart from Newcastle Station with a passenger service on 1 August 1934. The water crane is in position over the tender. (W.B. Greenfield, courtesy of the NELPG)

said. In reality, these tenders did little to shorten the actual journey times – but, again, it was all good publicity. Further corridor tenders were built at intervals until 1938 and eventually there was a total of twenty-two. At various times, they were coupled to engines of A1, A3, A4 and W1 classes. By the end of 1948, all of the corridor tenders were running with A4 class locomotives. The coal capacity was increased to 9 tons when the streamlined fairing was removed from the front of the top of the tender coal space. This was done after two A4 class locomotives ran out of coal working the up 'Coronation' at Hitchin. The use of the corridor tenders for the changing of crews on the move is superbly shown in the 1953 British Transport film 'Elizabethan Express', which depicts another London to Edinburgh non-stop train service.

On 1 May 1928 at 10.00 the Lord Mayor of London waved off the third of H.N. Gresley's 'Pacifics', No 4472 *Flying Scotsman*, and the train of the same name into the history books. No 4472 was one of five of H.N. Gresley's 'Pacifics' selected to work the prestigious non-stop 'Flying Scotsman' service from London to Edinburgh. Using the usual facility for water replenishment from the water trough system, these were able to travel the 392 miles between London and Edinburgh in 8 hours without stopping.

This was a record at this time for a scheduled passenger service, although the London, Midland & Scottish Railway (LMS) had attempted to upstage the event just four days earlier, with a one-off publicity run of its 'Royal Scot', a non-stop express run of the Euston to Edinburgh route of some 399.7 miles. It used an LMS compound-type locomotive, which was fitted with a modified tender.

The 1928, non-stop 'Flying Scotsman' express now had better catering and other on-board services for the travelling public, and could even boast an on-board cinema. Over time, things got even better for the weary railway passenger, as when the 'limited speed agreement' ended in 1932, the journey time was reduced to 7 hours 30 minutes, and by 1938 the time was 7 hours 20 minutes. In the British Railways era, the 'Flying Scotsman' service ceased to be a non-stop train and began stopping at Newcastle upon Tyne, York and Peterborough.

On 24 August 1935, No 4472 *Flying Scotsman* departs from Newcastle Station with an up passenger service. The station pilot, a J72 0-6-0 tank locomotive, simmers in the shadow of the station confines. (W.B. Greenfield, courtesy of the NELPG)

❮ In 1934, No 4472 *Flying Scotsman* achieved the first officially authenticated run of 100mph for a steam locomotive, whilst working a test train running down Stoke Bank. It is seen here at speed on that actual run on 30 November 1934. (Sir William McAlpine collection)

In the late 1950s, BR were committed to modernisation by using diesel locomotives and began devising a replacement for the Gresley designed 'Pacifics' on the East Coast Main Line. In 1962, the powerful 'Deltic' diesel locomotives took over from steam, with the result of a twenty-two-strong, Type 55 'Deltic' fleet. The 'Deltic'-hauled 'Flying Scotsman' service became the centrepiece of British Railways advertising, as had been the steam-hauled service for the LNER.

After the 'Deltics', the InterCity 125 High Speed Train, brought into service from 1976 onwards, was the regular provider of haulage on the East Coast Main Line's 'Flying Scotsman' service.

Because of the power of advertising and of the romantic nostalgic link with the past, since the privatisation of British Rail the 'Flying Scotsman' name has been maintained by the private operators of Anglo-Scottish trains on the East Coast Main Line. From 1996 to November 2007, the 'Flying Scotsman' service was operated by Great North Eastern Railway (GNER), which chose its name to bring up connotations of the LNER. It subtitled the East Coast Main Line route as 'The Route of the Flying Scotsman'. The 'Flying Scotsman' service was then operated by National Express East Coast until November 2009 and was operated seven days a week by its next operator East Coast Railways, a publicly operated company that had been created after the collapse of National Express East Coast.

Because of the persistent success of the 'Flying Scotsman' service, the train continues to run today as a major link between the capital cities of England and Scotland. In a direct response to the requirements of the business communities in Edinburgh and the North East, on 22 May 2011 East Coast started to operate its new early morning 'Flying Scotsman' service departing from Edinburgh at 05.40, with only Newcastle as an intermediate stopping point. This service has reduced the time taken to travel between Edinburgh and London King's Cross to just 4 hours!

2

WHO WAS HERBERT NIGEL GRESLEY?

Herbert Nigel Gresley was born in Edinburgh on 19 June 1876 and died on 5 April 1941, aged 65. He was truly a railway mechanical engineer of the highest order. From the beginning, his career brought him into contact with some of the great railwaymen of the day, such as Francis Webb, John Aspinall and George Jackson Churchward. H.N. Gresley established himself as every inch the equal of these great names and others of the past, and rose to become the GNR's carriage and wagon superintendent at Doncaster Works in 1905, moving on to become the locomotive engineer of the GNR, when Henry Albert Ivatt retired in 1911.

H.N. Gresley was the fourth son of the Revd Nigel Gresley, and although both of his parents were in fact English, he was born in Edinburgh, whilst his mother was in the Scottish capital to consult a gynaecologist. The family home lay at Netherseale, 4 miles south of Swadlincote in Derbyshire, which is a district associated with several generations of noted personalities with the name Gresley. He was educated at Marlborough College in Wiltshire, a school founded to educate the sons of the clergy.

His career started at that northern hotbed of the railways, Crewe, as a pupil to Francis Webb, the Chief Mechanical Engineer of the London & North Western Railway, which at that time was known as 'The Premier Line'. John Ramsbottom, Webb's predecessor, had established Crewe Works as the foremost engineering establishment in the world and had instigated the system of batch production, using components machined to such fine tolerances as to make them interchangeable. He also introduced the split cast-iron piston ring, which is still in use in every piston engine throughout the world today.

In 1860, John Ramsbottom developed the use of the water trough, from which locomotive tenders could be filled with water at speed, enabling long runs to be undertaken non-stop. H.N. Gresley would put this to good effect later in his career on *Flying Scotman*'s famous non-stop runs from London to Edinburgh.

When Francis Webb took up his position at Crewe on 1 October 1871, his first locomotives were similar to those of his predecessor, one of which, when rebuilt in 1887, was a 'Precedent' class 2-4-0 locomotive No 790 *Hardwicke*. It had two inside cylinders and associated valve gear. The locomotive had been named after the designer of the erstwhile Doric arch at Euston. No 790 *Hardwicke* is preserved today as part of the National Collection at York because of its achievements on 22 August 1895, when it took 2 hours and 6 minutes to travel the 141 miles from Crewe to Carlisle, at an average speed of 67.1mph. In doing so it set a new speed record during the 'Race to the North'.

Another London and North Western Railway (LNWR) class, the 'Teutonic', performed sterling duty in service. It had a 2-2-2-0 wheel

This beautifully executed line drawing of Patrick Stirling's 2-2-2 locomotive No 234 was completed by Nigel Gresley when he was just 13. (Author's collection)

Sir Nigel Gresley, one of Britain's most famous steam locomotive engineers, admiring his eponymous streamlined A4 class 'Pacific' locomotive No 4498, at King's Cross Depot on the day of the formal naming of the locomotive in 1937. (David Ward collection)

On 1 May 1976, the NRM-owned, former LNWR 2-4-0 No 790 *Hardwicke* piloted No 4472 *Flying Scotsman* on 'The Settle & Carlisle Centenary Train', which consisted of preserved passenger coaches and is seen here passing through Clapham, North Yorkshire. (Terry Robinson)

arrangement, with three cylinders and three sets of valve gear. But it had difficulties in starting since the driving wheels were not coupled and could even oppose each other. H.N. Gresley would have got a good understanding of the problems of the assembly and the setting up of the valve gear of these engines. The prospect of such complexities would have put him off the idea of having to use such a mechanism himself and would have been the foundation for his simple '2 to 1' lever mechanism, as was used on his 'Pacific' locomotives.

As seen earlier, when H.N. Gresley was at Crewe the great 'Race to the North' of the late nineteenth century was still raging. This battle had started seven years before between the East and the West Coast railway lines for the much-prized Edinburgh traffic. In 1888, the fastest journey time between London and Edinburgh had been reduced to 7 hours 27 minutes, but the competition had cooled down for a while.

However, with the opening of the Forth Railway Bridge in 1890, there was renewed competition to get even further north to Aberdeen and during 1894 the bridge had carried 26,451 passengers. During the night of 22 August 1895, the LNWR and the Caledonian Railway started to increase the speed of their services. They ran from London to Edinburgh in 512 minutes and beat their East Coast competitors' time by 8 minutes. The train used was very light (with a load of only 70 tons) and was worked by Webb's compound No 1309 *Adriatic*, which completed the first 158 miles to Crewe in 157 minutes. The locomotive was then changed for No 790 *Hardwicke*, which managed its record-breaking run to Carlisle – including surmounting the significant obstacle of the 900ft Shap Summit – in 126 minutes. The changeover of locomotives at Crewe took place at about 22.30 and was not unlike a Formula One pit-stop of today, with its highly choreographed and rehearsed actions. After the train had come to a

As it used to be in 1894! An image of third-class passenger accommodation, as was provided by the Great Northern Railway on the route to Scotland. (Author's collection)

Pennsylvania Railroad's K4s class 'Pacific' steam passenger locomotive No 3750 is one of only two survivors of a great fleet that once numbered 425. Here it awaits attention at the Railroad Museum of Pennsylvania in Strasburg in 2011. It was this type of locomotive that H.N. Gresley used as the basis for his A1 class 'Pacific' locomotives, of which *Flying Scotsman* was the third example to be completed. (Author)

complete halt, the locomotive was then immediately backed up to slacken the couplings between it and train, which were then hastily uncoupled. The locomotive was then moved away sharply over the points to get out of the way to allow the small *Hardwicke* to reverse back on to the train. The coupling was smartly hooked on to the first carriage, the screw tightened up and only 90 seconds after the manoeuvre began, the train was ready to depart. The other stops on the journey were also quick, with that at Carlisle taking 3 minutes and that at Perth 5 minutes.

A different event of which H.N. Gresley would also have been aware occurred in 1895, when another of Webb's compounds, No 1305 *Ionic*, ran the 299 miles from Euston to Carlisle non-stop. This was the longest distance permitted anywhere for a non-stop run and was a world record until 27 April 1928, when H.N. Gresley's A1 class 'Pacific' No 4472 *Flying Scotsman* ran non-stop from London to Edinburgh during a proving run for the non-stop service.

In 1898, H.N. Gresley left Crewe and moved on to study under John Aspinall at the Horwich Works of the Lancashire & Yorkshire Railway (L&YR), where Aspinall was responsible for some very fine locomotives and where he also introduced the first attempt at superheating the steam in a boiler. It was at Horwich that H.N. Gresley discovered that L&YR express locomotives had inside cylinders and inside valve gear. At that time, the inside cylinder arrangement was favoured in the UK, as the cylinders were placed directly under the smoke box. This was popular as it made for short steam and exhaust pipes; also, the cylinders formed a neat tie between the frames, whereas outside cylinders imposed a greater stress on the frames and didn't look so neat in their overall appearance.

Shortly after Nigel Gresley had arrived at Horwich, the L&YR Railway brought out a new 4-4-2 or 'Atlantic'-type locomotive, the type of which was first used by the Atlantic Coast Line in the USA, hence the name. The small trailing wheels allowed the use of a Wootten-designed firebox, which was wider than normal, and this is probably where Gresley got the idea to use a wide firebox on his 'Pacific'-type locomotives.

As the L&YR were bringing out their 'Atlantics', the GNR's Chief Mechanical Engineer, H.A. Ivatt, who had worked under John Aspinall in Ireland, was also producing his own design of 'Atlantic'-type locomotives and just managed to get out his first one, No 990, from Doncaster a few months ahead of the L&YR. It was named *Henry Oakley* after the GNR's chairman and was one of the few locomotives to be named by the GNR. Together with No 251, a larger version produced in 1902, these 'Atlantics' were the 'first-line' locomotives of the GNR when Nigel Gresley took over

^ During June and July 1923, two of Sir Vincent Raven's 'Pacifics' were completed at Darlington's North Road Works, and with twelve examples of H.N. Gresley's 'Pacifics' soon to be completed, a total of fourteen 'Pacifics' would soon be available for service. This prompted H.N. Gresley to initiate comparative trials between the two types of locomotive, using his locomotive No 1472 and Sir Vincent Raven's No 2400, later to be named *City of Newcastle*, which is seen preparing to depart from King's Cross Station in June 1923 on one of these runs. (Author's collection)

^ The Great Central Railway initiated the design of a four-cylinder 2-6-0 + 0-6-2 'Garratt' locomotive, which evolved into a 2-8-0 + 0-8-2 configuration, with the design modified by H.N. Gresley in 1924 to use three cylinders at each end. Although there were plans for two such locomotives, only one was built. This lone engine, No 2395, was built by Beyer Peacock in Manchester and was delivered to the LNER ready to be displayed at the Stockton & Darlington centenary celebrations on 1 July 1925. (Author's collection)

in 1911. No 251 and No 990 have both been preserved and are part of the National Collection at York.

In 1901, H.N. Gresley spent the summer season at Blackpool's locomotive running shed and would have been very busy there, for in the heyday of that great British seaside resort it wasn't uncommon for 100,000 people to be transported to and from Blackpool Station on a single summer's day. Blackpool, in fact, was where Nigel Gresley met his future wife. They married in 1901, within the same year that they had first met.

During 1901, H.N. Gresley moved to Newton Heath, where three years later he was appointed assistant superintendent of the Carriage and Wagon Works, where his employment lasted just one year. In February 1905, just a few months short of his twenty-ninth birthday, he was appointed by the GNR as the carriage and wagon superintendent at Doncaster Works.

Along with the LNWR, the GNR was the one of the most prestigious lines to run north out of London. The GNR had been running the predecessor to the 'Flying Scotsman' service in conjunction with the North Eastern and North British Railways since 1862; as we have seen, during this period it was known as the 'Special Scotch Express'.

One of H.N. Gresley's first tasks in his new position at the Carriage and Wagon Works was to eliminate the use of gas as a source of heating in passenger vehicles, due to the risk of fire spreading in the event of a collision or derailment of the vehicle.

H.N. Gresley's first full design at the Carriage and Wagon Works consisted of two steam-powered rail-motor cars, built in 1905. In the design, a steam-powered bogie unit which had two outside cylinders used valve gear driven by an external Walschaerts mechanism, the first time this was ever used by the GNR. The bogie was coupled to a new-style coach with an elliptical roof. This all made a big impression on H.N. Gresley, as it would have made it much easier to adjust and maintain the mechanics of the vehicle than the inside type that he had been used to working with. He subsequently used the design on most of his locomotives.

In 1906, H.N. Gresley produced a prototype bogie luggage van, No 126, for the East Coast Joint Stock, which was the first vehicle to be built at Doncaster with an all-steel under frame. The body sides were of teak, which must have been deemed successful as it became the standard finish for future LNER main-line stock, including the subsequent 'Flying Scotsman' train.

In 1907, H.N. Gresley produced some articulated coaches for both suburban and main-line use which proved to be very successful; this form of construction was later used in his high speed trains as a means of saving weight. This method of articulating bogies was subsequently incorporated into the design for Eurostar trains, as used on Channel Tunnel rail services from St Pancras International to Europe.

H.N. Gresley also produced some close-coupled double-brake vans, which allowed the weight of freight trains to be increased to 600 tons and which later induced him to increase the output from his freight locomotives and ultimately to adopt a three-cylinder design.

In 1908, H.N. Gresley adopted, as a standard bogie, a design that had been produced by Spencer, Moulton & Company. This design became known as the standard LNER Gresley bogie and came close to being adopted as the standard type for British Railways carriages after nationalisation took place in 1948. By 1910, Gresley had developed the large-boilered 'Atlantics', fitting them with 'superheating' capabilities – where generated steam is given an extra pass through the heat generated from the firebox, which boosts the performance of the engines and allows them to be capable of fantastic feats of haulage.

When H.A. Ivatt retired in 1911, Nigel Gresley – still only 35 years old at the time – was appointed as the Locomotive Engineer of the GNR and took on the responsibility for the provision of all the rolling stock and the necessary motive power to keep this railway operational. Considering his portfolio, and given he was effectively fulfilling the same role as the outgoing H.A. Ivatt, it may seem strange that the GNR didn't accord him Ivatt's title of Chief Mechanical Engineer. Gresley's intention had been to produce an engine capable of handling, without assistance, main-line express passenger services that were fast reaching the limits of the hauling capacity of H.A. Ivatt's large-boilered 'Atlantic' 4-4-0s.

H.N. Gresley's initial plan became the 'Pacific' project of 1915, for an elongated version of the Ivatt 'Atlantic' design, but using four cylinders. He realised that he needed inspiration from somewhere, so took as his model the new Pennsylvania Railroad K4s class two-cylinder 'Pacific' of 1914 (the 's' indicated that the locomotive was superheated). This class had itself been updated from a series of prototypes scientifically developed in 1910, under Francis J. Cole, Alco's Chief Consulting Engineer at Schenectady, as well as the Pennsylvania Railroad's K29 class (Alco's prototype of 1911), also designed by Cole. Descriptions of those locomotives appeared in the British technical press during this time and they gave H.N. Gresley some of the elements necessary to design a thoroughly up-to-date locomotive.

Unlike the K4s class locomotive, Gresley's firebox was not of the flat-topped Belpaire variety, but was of a round-topped design that was in line with GNR tradition of the time. However, there were features in common with the American K4 type: the downward profile towards the

^ On 20 October 1936, No 4472 *Flying Scotsman* left Doncaster Works after a non-classified repair. Corridor tender No 5324 had been replaced with a GNR design No 5290. Here *Flying Scotsman* is seen at the works waiting to be transferred to Carr loco depot to re-enter traffic. (Sir William McAlpine collection)

back of the firebox; the boiler tapering towards the front; heat transfer and the flow of gases helped by the use of a combustion chamber extending forward from the firebox space into the boiler barrel; a boiler tube length limited to 19ft; and the boiler pressure was rated at 180psi. But for all that H.N. Gresley must have learned from American practice, he made no subsequent reference to the Pennsylvania's 'Pacifics', with regards to the design of his own A1 class 'Pacific'.

In 1920, Nigel Gresley's work to reorganise Doncaster Works for the production of munitions during the First World War was recognised, when he was rewarded with a CBE. Then in April 1922, No 1470, subsequently named *Great Northern*, became the first of H.N. Gresley's 'Pacific' locomotives to enter service. The second, No 1471 *Sir Frederick Banbury*, entered service in July 1922, rapidly followed by an order for a second batch of ten locomotives to be constructed. The design was deemed by the GNR Board of Directors to be a revolutionary and immediate success.

H.N. Gresley's 1470 class 'Pacific'-type locomotive design was the third GNR locomotive type to incorporate his universal three-cylinder layout. With this system, drive from all three cylinders is concentrated on the

No 4472 *Flying Scotsman* is seen working a heavy express passenger service through Low Fell, 2½ miles south of Gateshead, during the mid-1930s. (W.B. Greenfield, courtesy of the NELPG)

The world speed record for steam traction was captured by the LNER's A3 class locomotive No 2750 *Papyrus* on 5 March 1935. On this day it reached a top speed of 108mph on a high-speed test run between Newcastle and London. (David Ward collection)

middle coupled axle using cranks set at 120 degrees and using the Gresley conjugated valve gear system, with the motion of the two outside valve spindles being transferred to the middle cylinder, eliminating the need for an inaccessible middle set of valve gear between the frames.

Sir Vincent Raven, the Chief Mechanical Engineer for the NER, had also designed a 'Pacific'-type locomotive at about the same time. The NER's locomotive had hastily been constructed to have it in use just before the amalgamation in 1923 of the constituent railway companies which became the L&NER. The principal companies were: the Great Eastern Railway; Great Central Railway; Great Northern Railway; Great North of Scotland Railway; Hull & Barnsley Railway; North British Railway; and North Eastern Railway. Along with the change of ownership came a change of chief engineer. John Robinson, of the GCR, was offered the position with the LNER, but declined and the post went to the former GNR's locomotive engineer – one Nigel Gresley. The new LNER was the second largest of the 'Big Four' railway companies in Britain.

In 1923, the third 'Pacific' locomotive of the 1470 class to be built, No 1472, became the first locomotive to be turned out for the newly formed L&NER, as it was then called, even though it had been ordered during GNR days in 1922. Realising the need for standardisation, Nigel Gresley made comparative trials with one of the NER-built 'Pacific' locomotives. In 1923 the newly formed LNER conducted a series of tests comparing these two new designs. Raven's 'Pacific' No 2400 had no problem producing enough steam, but Gresley's 'Pacific' No 1472's boiler pressure varied during the tests, whilst No 1472 consumed notably less coal and water. After the comparative test data had been received and analysed, the choice adopted as the standard express passenger locomotive for the LNER main line was Nigel Gresley's GNR designed 'Pacific' design locomotive. This was given the class designation A1 within the LNER's locomotive classification system, with the NER's design being classified as A2. The Gresley 'Pacific' was considered to be the better design, partly because it was considered to be more amenable to improvement in the future.

On 6 February 1924 an instruction was issued, ordering '... that 3000 be added to all existing GNR locomotive numbers', and so *Flying Scotsman*, as it was to become, was renumbered as No 4472 and displayed at the British Empire Exhibition at Wembley, along with the first member of the GWR's 'Castle' class, No 4073 *Caerphilly Castle*, which was the first locomotive to be constructed at Swindon Works after the First World War.

The 'Castle' class was claimed by the GWR to be the most powerful locomotive in Britain, with a tractive effort rated at 31,825lb. This was in spite of the fact that it weighed in at 19.6 tons less than the

Gresley-designed 'Pacific'. LNER contested this claim, and eventually the two railway companies ran comparative exchange trials between the two types of locomotive, from which the GWR emerged triumphant with its locomotive No 4079 *Pendennis Castle*.

The LNER learned valuable lessons from the trials, which resulted in a series of modifications being carried out from 1926 on No 4477 *Gay Crusader*. Changes to the valve gear included increased lap and longer travel, in accordance with GWR practice. This allowed for fuller exploitation of the expansive properties of steam, whilst reducing back pressure from the exhaust, transforming performance and improving economy. The economies in coal and water consumption achieved were such that the 180psi 'Pacifics' could now undertake long-distance non-stop runs that had previously been impossible.

Between June 1924 and July 1925, another forty A1 class locomotives were produced. Then, under H.N. Gresley, there followed a complete redesign of the valve gear on the A1 class locomotives, a change which was applied to the whole class. Locomotives with modified valve gear had a slightly raised running plate over the cylinders, in order to give more room for the longer conjugation lever necessary for the longer valve travel.

H.N. Gresley was dissatisfied with the lower superheating temperatures of up to 600°F in his 'Pacifics', compared with up to 800°F in the GNR 'Atlantics', due to the greater cooling effect on the tube elements caused by the longer boiler on the 'Pacific' design. One 'Pacific' was fitted with a 'Type E' superheater, but this failed to produce any significant results. H.N. Gresley therefore decided to increase the number of flues in the Robinson superheater, which was already in use in the A1s. At the same time, it was also decided to increase the boiler pressure from 180psi to 220psi. The first five of these new boilers were ordered in 1927 and by the end of 1927, No 2544 *Lemberg* and No 4480 *Enterprise* had been rebuilt with the new system. These were followed by three more rebuilds in 1928. The cylinder size was kept the same at 20in, except for No 2544 which had cylinders lined to 18.25in, so that the tractive effort would be similar to the A1s. This allowed comparative trials to be made and they were deemed to be a success. The modified locomotives were given the designation of A3 class and twenty-seven further A1s were rebuilt as A3s between 1928 and 1935. These rebuilds had slightly smaller cylinders of 19in and the original five rebuilds were fitted with this new size.

Due to the wider 'superheater header' on the A3 class boilers, the ends projected slightly through the sides of the smoke box, hence cover plates were fitted to cover up these projections. These cover plates are the main external distinguishing feature between the A3s and the A1s.

During this period, Nigel Gresley lived at Salisbury Hall, near St Albans in Hertfordshire. In the moat around his home, Gresley developed an interest in breeding wild birds and ducks and, appropriately enough, amongst the species were mallards. Perhaps they inspired the naming of the most famous of Gresley's later A4 class? The hall still exists today as the home of the de Havilland Aircraft Heritage Centre.

The acceleration of the non-stop 'Flying Scotsman' service in 1932 from 8¼ to 7½ hours was an indication of changing railway management policy, with a greater emphasis on speed as an incentive to passenger travel, especially during a period of trade depression and increasing competition from road transport. So the exciting news from Germany of 'The Flying Hamburger' service being worked by a two-car diesel-electric articulated unit, working speeds of up to 100mph, led H.N. Gresley to go and visit it in action. He concluded that, with a lighter train, his 'Pacifics' would be capable of a similar achievement. So, in the course of a trial run on 30 November 1934, No 4472 *Flying Scotsman*, in the capable hands of Driver Sparshatt and Fireman Webster, hauling a four-coach train, including the dynamometer car and a restaurant car, covered the 186.8 miles from King's Cross to Leeds at an average speed of 73mph. On the return trip, and with an additional two coaches in the formation, No 4472 *Flying Scotsman* achieved 100mph descending Stoke Bank, becoming the first steam locomotive anywhere in the world to be officially authenticated as having achieved a speed of 100mph.

As well as rebuilding his A1 class locomotives to become the A3 class, H.N. Gresley built a series of new A3 class locomotives. On 5 March 1935, A3 class 'Pacific' No 2750 *Papyrus*, built at Doncaster in 1929, passed Little Bytham at 106mph as it was descending Stoke Bank with a six-coach train weighing 217 tons gross. It reached a maximum speed of 108mph soon afterwards and was noted as travelling at 102mph through Essendine. As the end of steam in the UK approached, *Papyrus* was withdrawn from service on 9 September 1963 and was subsequently scrapped.

In September 1935, H.N. Gresley introduced his A4 class streamlined locomotives to haul a new express train called the 'Silver Jubilee', which was planned to run between London King's Cross and Newcastle. The new service was named in celebration of the twenty-fifth year of King George V's reign. The LNER had authorised Nigel Gresley to produce a streamlined development of the A3 class. Initially four locomotives were built, all with the word Silver as part of their names. The first of these was No 2509 *Silver Link*, which entered service on 7 September 1935. The other three were No 2510 *Quicksilver*, No 2511 *Silver King* and No 2512 *Silver Fox*. During a press run to publicise the service, *Silver Link* twice achieved a speed of 112.5mph, breaking the British speed record, and sustained an average speed of 100mph over a distance of 43 miles.

Following the commercial success of the 'Silver Jubilee' train, specially built A4s were assigned to other express services, to make best use of the streamlined engines. The 'Coronation' began running from London to Edinburgh in July 1937, while four months later the 'West Riding Limited' started up from Bradford and Leeds to London.

The A4s' streamlined design gave them high-speed capability as well as making them instantly recognisable, and one of the class, No 4468 *Mallard*, achieved the official record as 'The Fastest Steam Locomotive in the World'. Thirty-five of the class were built in total to haul express passenger trains on the East Coast Main Line from King's Cross to Edinburgh via York and Newcastle. They remained in service on the East Coast Main Line until the early 1960s when they were replaced by 'Deltic' diesel-electric locomotives. Several A4s saw out their remaining days until 1966 in Scotland, particularly on the Aberdeen to Glasgow 3-hour express trains, for which they were used to reduce the duration from 3½ hours.

In 1936, Herbert Nigel Gresley was appointed an honorary Doctor of Science by Manchester University. The following year, he received perhaps the ultimate honour in being knighted by King Edward VIII in recognition of his work in the railway industry.

Herbert Nigel Gresley served on several government-appointed committees, including those considering automatic train control and the electrification of the railways. He was President of the Institution of Mechanical Engineers in 1935 and was twice President of the Institution of Locomotive Engineers, in 1927–28 and 1934–35.

As well as his pioneering locomotive designs, another of his major achievements was the establishment of a locomotive testing station in the UK. He had long believed this to be of great importance to locomotive engineering in the country and his efforts resulted in a national testing centre being constructed jointly by the LNER and the LMS at Rugby. Work had commenced in 1937, but was postponed on the outbreak of the Second World War and unfortunately Nigel Gresley didn't live to see its completion. He died on 5 April 1941 at his home at Watton-at-Stone at the age of 65. Four days later he was buried in Netherseal, Derbyshire. On the same day a memorial service was held at Chelsea Old Church, at which his professional and business friends were present. This was during the Blitz on London and the following week the church was severely damaged by bombing.

Nigel Gresley designed some of the most famous steam locomotives in the world, including the A1, A3 and A4 classes, and his engines are considered by many to be elegant, both aesthetically and mechanically. His invention of a three-cylinder design with only two sets of Walschaerts valve gear, coupled with his conjugated 2 to 1 valve gear, produced smooth running and power at a lower cost than would have been achieved with the more conventional three sets of Walschaerts gear. His amazing innovations enabled his locomotive No 4472 *Flying Scotsman* to become the first steam locomotive in the world to be officially authenticated at attaining the speed of 100mph, and his A4 class locomotive, No 4468 *Mallard*, still holds the record for being the fastest steam locomotive in the world, having reached the amazing speed of 126mph.

3

THE GREAT NORTHERN RAILWAY YEARS

By the 1840s, it was possible to travel from Yorkshire to London by rail, but the journey was long and circuitous via Derby and Rugby to Euston. The London & York Railway (L&YR) proposed that a route be constructed from London heading directly north and passing through Peterborough, Grantham and Doncaster. There would be a loop line from Peterborough to Boston, Lincoln and Doncaster and there would be branches to Sheffield and Wakefield.

The renowned surveyor and civil engineer Joseph Locke performed the initial survey, but his proposal met with very stiff opposition from a group led by George Hudson, the so-called 'Railway King'. With Hudson employing every kind of delaying tactic imaginable, Locke was put under so much pressure that he resigned his commission at very short notice. But upon receipt of the resignation, another renowned civil engineer, William Cubitt, was employed in a matter of a few days. Once the bill was passed by parliament, the 'London & York' name was dropped and the new title of the Great Northern Railway (GNR) was adopted in its place.

In December 1848, plans for the station at King's Cross were initially made, and the main portion of the station, which today includes platforms 1 to 8, was opened on 14 October 1852. This replaced a temporary terminus that had been in use since August 1850.

Previously, the area of King's Cross was a village named Battle Bridge, situated next to an ancient crossing of the River Fleet. The name of this village arose from the tradition that it was the site of a battle between Boudica and the Romans, supposedly her final battle, and indeed folklore states that the warrior queen herself is buried between platforms 9 and 10 of King's Cross station.

As train services started, the earliest locomotives ordered for the GNR were of standard types built by the well-known makers of the time. The late arrival of the GNR on the scene, however, gave it an advantage as standard designs had evolved and had been improved. For example Order No 198, of 5 February 1847, covered six locomotives and a further forty-four were covered by Order No 203, dated 4 March 1847. These small locos had 5ft 6in driving wheels, with boilers working at the relatively low pressure of 90psi and were taken into stock between 1847 and 1850. Archibald Sturrock, having previously been Daniel Gooch's assistant on the GWR from 1840, became the locomotive superintendent of the GNR in 1850 and during his sixteen years there designed some amazing locomotives. On these 'bought-in' primitive engines he fitted compensating levers, which assisted them to run on the relatively poor and lightweight track.

Progress improved the design of steam locomotives forward, leading to a most notable locomotive on express duties – Patrick Stirling's 4-2-2

An 1852 view of King's Cross Station just after opening. Designed by Lewis Cubitt and built by builders John and William Jay, it was the southern terminal of the GNR. The roof spans, which are 105ft wide by 800ft long, were originally supported with ribs of laminated timber, which were subsequently replaced with steel. (Author's collection)

'Single' class, a locomotive with 8ft 1in driving wheels. Patrick Stirling had designed his original 'Single' in 1870 at a time that new engines were required for the many new routes that were still being built. The 'Singles' were built for speed, power and to handle the continuous gradients on the main GNR line from London to York, and were also built to compete against the Midland Railway and L&NWR in the 'Races to the North'. Subsequently, H.N. Gresley adapted the 'steam collection mechanism' from Stirling's 'Single' locomotives when he fitted 'banjo domes' (basically an elongated form of steam collector developed to help prevent priming) on his improved A1 class which became the A3 class 'Pacifics'. The Stirling 'Single', when rebuilt by H.A. Ivatt, represented the penultimate development of single-drive-wheeled locomotives.

Over this period, train weights continued to rapidly increase and until the end of the nineteenth century, trains such as the 'Flying Scotsman' (or 'The Special Scotch Express' as it was still officially known) included six-wheel coaches in their formation. But in 1900, a revolutionary change took place, which was the introduction of new trains of American design and appearance. Composed of eight bow-ended, 65ft 6in-long, twelve-wheel cars, including restaurant cars, they weighed in total some 265 tons. Similar new coaching stock arrived on other Anglo-Scottish trains and on the London to Leeds service, so increasing train loads generally was becoming the norm. Continuing development, H.A. Ivatt improved the water-boiling ability of his engines, with amazing results.

With Ivatt having replaced the highly respected Patrick Stirling after the latter's death in office, he had a lot to live up to in maintaining both the working quality and aesthetic artistry of Stirling's locomotives – particularly the Stirling 'Single', which was considered the ultimate in British locomotive artistry. As, by the turn of the twentieth century, greater power and adhesion were being sought, Ivatt drew inspiration from the Baldwin Locomotive Works design for the Atlantic Coast Line in North America. By 1897 he had designed the first 4-4-2 or 'Atlantic' type of locomotive to be in service in Britain, and a year later Britain's first 4-2-2 'Atlantic' locomotive, No 990, was ready for use.

The new 'Atlantic' design was in some ways a complete reversal of a predecessor's policy and was different in that it involved more than just increasing the proportions embodied in an engine, as had been the case

This magnificent model is of one of Patrick Stirling's famous 'eight-footer' locomotives. The most noticeable characteristic was the huge 8ft driving wheels, which allowed them to reach very high speeds. The surviving member, No 1, is part of the National Collection at York. (Author's collection)

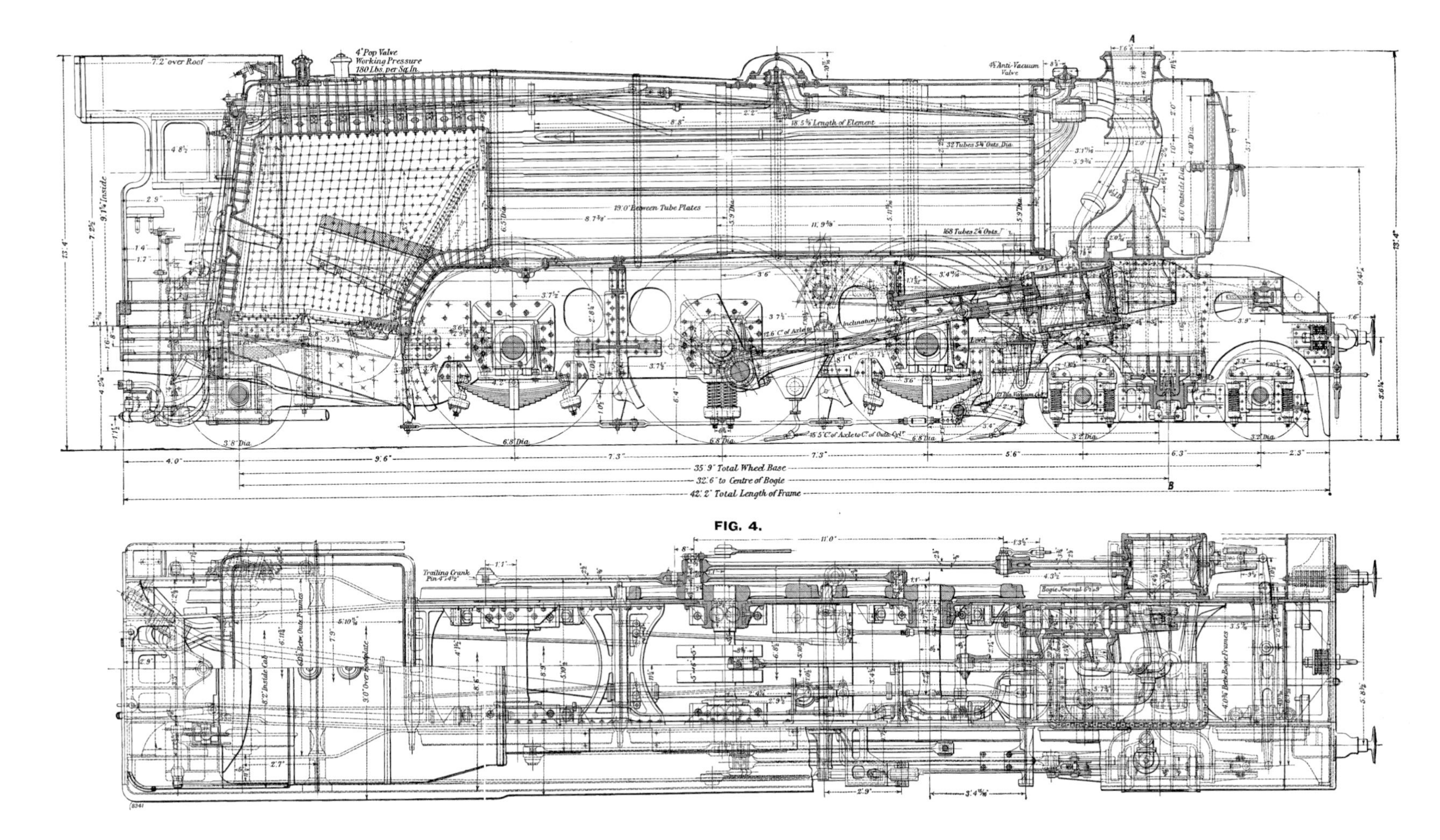

^ A beautifully detailed drawing showing the general layout of a Gresley-designed A1 class 'Pacific' locomotive of the early 1920s. (Author's collection)

with previous design improvements. Instead of 19½in by 28in cylinders, there were none bigger than 18½in diameter by 24 stroke; instead of a boiler with no more than 1,032 sq. ft heating surface and a fire-grate area of 20 sq. ft, the new 4-2-2 had a heating surface of 1,442 sq. ft and a fire-grate area of 26.8 sq. ft. The odd result was that H.A. Ivatt's 'Atlantic' had a nominal tractive effort of 15,860lb, whereas that of Stirling's last 4-2-2's was 16,100lb. But the tractive force formula means nothing unless there is sufficient steam to make it effective. Ivatt had publicly proclaimed that the capacity of a locomotive was 'its ability to boil water', and here was the principle in action. There were no questions, of course, as to which was actually the more powerful of the two types. In total, twenty-two examples were built between 1898 and 1903 at Doncaster Works. The large capacity boiler gave extra steam-raising capacity, which gave the 'Atlantics' the edge over Stirling's single wheelers. The first production 'Atlantics' entered service in 1900 and were fast, lively runners, to the extent that H.A. Ivatt had to formally caution his drivers to rein in the speed due to uneven stretches of track between London and Doncaster which were considered unsafe for high speed running! But, in turn, the engine men would have told H.A. Ivatt that the cylinders were no match for the boiler. To achieve the expected performance these first 'Atlantics' had to be worked at undesirable and uneconomic rates – to use a term of the day, the engines were 'thrashed'.

The 'Atlantics' were the first GNR engines to be built with a wide firebox, directly as a development from the original 'Atlantics'. No 251 became symbolic of the GNR and featured in most of the GNR's advertising and even on timetable covers. The 'Atlantics' were acknowledged as superb locomotives on the East Coast express trains and were only surpassed by the larger and more powerful 'Pacifics' such as A3 class *Flying Scotsman* and A4 class *Mallard*.

When H.N. Gresley was appointed the locomotive engineer of the GNR, he quickly began to think about large express passenger locomotives. His predecessor, H.A. Ivatt, had already experimented with compound four-cylinder 'Atlantics' with high-pressure boilers and wide fire-grate areas and, indeed, H.N. Gresley's first designs followed these ideas, so in 1915, an 'Atlantic' was suitably modified. There then followed plans for two different 'Pacific' designs, one of which was simply an extended version of his new 'Atlantic' design. Around this time, H.N. Gresley became interested in three-cylinder designs and produced his famous conjugated valve gear. This mechanism worked whereby the operation of the valve for the middle cylinder was derived from the motion of the other two cylinders. Although the design was patented in November 1915, H.N. Gresley readily admitted that he used a lapsed patent of Harold Holcroft's as inspiration. The First World War caused express passenger designs to be shelved; however, H.N. Gresley did test his new valve gear design with a three-cylinder 2-8-0 locomotive, which was completed in 1918. Following this was a much-needed express goods 2-6-0 locomotive, which used a much simpler version of the conjugated gear.

With the war behind him, in 1920 H.N. Gresley set about developing his plans for an express passenger 'Pacific' with gusto, incorporating his prized conjugated valve gear. The K3 class previously designed with the same valve gear exhibited valve over-run at high speeds, and so the maximum travel was reduced, with a resulting reduction in the locomotive's performance.

Primarily, H.N. Gresley's 'Pacific' design was inspired by the Pennsylvania Railroad's K4s class 'Pacific' design. The boiler was reduced to meet the GNR loading gauge, but kept to broadly the same tapered shape; the firebox was rounded as was the case in current GNR practice and not square-topped as in the American version; and it used tubes less than 19ft long. So by combining the K4s design with many of his original ideas, a formal design was put together.

On 10 January 1921 the GNR issued engine order No 293, giving Doncaster Works authority to construct two of H.N. Gresley's designed

^ One of the first pictures taken of No 1472, even before it was named *Flying Scotsman*. (Sir William McAlpine collection)

'Pacific' tender locomotives. Interestingly, just two months later, authorisation was given by the NER to construct two of Sir Vincent Raven's 'Pacific' tender locomotives at Darlington.

By 30 March the following year, Nigel Gresley had put the first of his three-cylinder 'Pacific' locomotives to work on the GNR, where it became the first serious and practical 4-6-2 'Pacific' to be used in the UK. The GWR had, in 1908, built a solitary example, No 111 *Great Bear*, to satisfy demands from the directors for the largest locomotive in Britain. This was considered the company's flagship engine; however, in service it wasn't a significant improvement on existing classes, and had highly restricted route availability. It was subsequently rebuilt as a 4-6-0 locomotive.

After the authorisation to build the first two NER 'Pacifics' was received at Darlington on 30 March 1922, line drawings were published in July 1922. It was thought that they were timed to counter the publicity of the appearance of the GNR's first 'Pacific' – a petty attempt to steal GNR's thunder, some thought. The two NER locomotives were recorded as being delivered by December 1922, just in time for the 'Grouping' in 1923, even though No 2400 was still in paint-shop grey undercoat and No 2401 didn't run until January 1923.

They had the nickname of 'skittle alleys', because of the great length of their parallel boilers. Like H.N. Gresley had done with his prototype, Sir Vincent Raven's design of 'Pacific' was obtained by stretching his own highly successful three-cylinder 'Atlantic' design, the C7 class, which had monopolised the East Coast Main Line between York and Edinburgh. He increased the cylinder size and the boiler diameter, while shortening the grate slightly. A wide firebox was fitted which spread out over the trailing wheels, and as this had to be behind the driving wheels, a further increase in length was inevitable, which proved to be a significant restriction on route availability. To counteract the adverse effects of longer boiler tubes, some of the tubes were increased in diameter and the boiler pressure was increased from 175psi to 200psi.

The three-cylinder drive, which worked on to the front coupled axle was a left over from the C7 design. One of Sir Vincent Raven's draughtsmen, Dick Innes, recorded in his diary that he had pointed out to Raven that 'the "Pacific" was a long engine and that if the drive was to be to the first pair of coupled wheels, then the locomotive would be even longer ...' Raven was reputedly not impressed with this remark; losing his temper and banging the table, he shouted '... I have always driven on to the leading coupled axle and I am going to do it with the "Pacific"!' The problem was that this configuration required that six eccentrics and the middle big end would need to be accommodated on the crank axle – a difficult engineering achievement, causing many fabrication problems. An 'eccentric' is a disc solidly fixed to the rotating axle of the driving wheels with an offset centre – it therefore has an 'eccentric' rotation. Its purpose is to convert the linear reciprocating motion from the piston and cylinder into a rotary action in order to turn the driving wheels. A similar arrangement was to be found on the GNR 'Atlantics'. With the valve arrangements and steam passages unchanged from the 'Atlantics', this proved to be a handicap to the required handling of large volumes of steam. Sir Vincent used three sets of independent Stephenson valve gear, which also had to be fitted in a confined space and which resulted in very limited bearing sizes. This was perhaps acceptable for the C7, but the bearings on the Raven 'Pacific' valve gear were expected to handle significantly more power. This was all compared with H.N. Gresley's much simpler designed conjugated valve gear for the middle piston, which was considerably smaller and needed far less maintenance.

Although the NER's 'Pacifics' were the biggest engine type that they had built, they did have a number of problems and few could claim it was their best. General opinion at the time held that the Raven 'Pacifics' were rushed out, so that they would be completed before the NER ceased to exist. The results obtained with the first two engines were reasonable and three more were ordered in February 1923. Before construction began, H.N. Gresley modified the rear bogie to use his preferred 'Pacific' trailing axle design (also known as a Cartazzi axle).

There was little difference in performance between the Gresley and Raven design 'Pacifics' when they were tested against each other in 1923. With the exception of a few visits to King's Cross in 1923 for trials, the Raven design 'Pacifics' spent their first ten years between Grantham and Edinburgh, with some runs to Leeds, and were originally allocated to Gateshead, before being moved to York in 1934 where they mainly ran heavy secondary express passenger trains on the East Coast Main Line, although they also occasionally ran freight and mail trains. Class example No 2402 was the first locomotive built by the LNER to be scrapped.

So, the first two of the GNR's 'Pacific' locomotives appeared as No 1470 and No 1471 and subsequently became classified as A1 class locomotives. They were very different from the existing GNR 'Atlantics' and caused quite a sensation, provoking great interest when they first arrived. As well as the very simple 2 to 1 conjugated valve gear, they had many other innovative features, including large cabs and, for the first time in locomotive history, the footplate crew were provided with padded seats.

4

THE LONDON & NORTH EASTERN RAILWAY YEARS

The grouping of the railways of Britain in 1923 was intended to stem the losses being made by many of the country's 120 railway companies. The amalgamation of Britain's railways resulted in four large companies: the Great Western Railway (GWR); the London, Midland & Scottish Railway (LMS); the Southern Railway (SR); and the London & North Eastern Railway (originally shown as L&NER and subsequently becoming the LNER). The process of grouping took place on 1 January 1923. The L&NER was the second largest of the 'Big Four' railway companies and had a total route mileage of 6,590 miles, which it worked until nationalisation took place on 1 January 1948.

The LNER owned:

7,700 locomotives
20,000 coaching vehicles
29,700 freight vehicles
140 items of electric rolling stock
6 electric locomotives
10 rail motor cars
6 turbine steamers
36 other steamers
Assorted river boats and lake steamers

Due to the success of H.N. Gresley's 'Pacifics' No 1470 and No 1471, which had been completed in 1922, a further order of ten 'Pacific' locomotives had been authorised by the GNR. The first locomotive to be constructed in this batch subsequently became the first express passenger locomotive to be completed for the newly formed L&NER. It was given works number 1564 and running number 1472 and left Doncaster Works on 7 February 1923.

The as-yet-unnamed No 1472, which also had the honour of being the first L&NER locomotive to be turned out in that company's new standard apple-green livery, had three 20in by 26in cylinders and was equipped with boiler No 7693, which had been set at a working pressure of 180psi. It was coupled to a GNR-designed eight-wheeled tender, No 5223, which had the letters L&NER painted on its tank sides above the number 1472.

On 22 February 1923, No 1472 was displayed at London Marylebone Station to be admired by the L&NER Board of Directors and members of the public. Two days later No 1472 was allocated to its first depot, Doncaster, and entered service.

With two Raven-designed 'Pacifics' already completed, together with the twelve examples of H.N. Gresley-designed 'Pacifics', a total of fourteen 'Pacific' locomotives would soon become available for service. H.N. Gresley decided a comparison of the two types was in order. Using Sir Vincent

❮ No 4472 *Flying Scotsman* after it had been prepared at Doncaster Works for exhibition at the British Empire Exhibition, Wembley in 1924. Notice the brass edging to the wheel splashers and the burnished wheel rims. (Author's collection)

❮ No 4472 *Flying Scotsman* is seen at Doncaster Works after it had been wrapped up in a 'hessian cover' to preserve its immaculate finish, just prior to its departure for Wembley, where it was exhibited at the British Empire Exhibition during 1924. *Flying Scotsman* had been wrapped up by the grandfather of Peter N. Townend, the pragmatic 'shed-master' of King's Cross 'Top Shed' locomotive depot. (Sir William McAlpine collection)

Raven's locomotive No 2400, later to be named *City of Newcastle*, and the third of H.N. Gresley's locomotives, No 1472, the scene was set for a series of comparative trials between the two locomotive types.

H.N. Gresley used a dynamometer car which recorded data on moving rolls of paper ensuring scientific analysis of the trials. Working the hardest 'turns' on the timetable at that time, each locomotive operated three return journeys between Doncaster and London, the 10.51 Doncaster to King's Cross, plus the return working (the 17.40 departure).

The doyen of the Gateshead's drivers, Tom Blades, was at the regulator of No 2400, which consumed coal at the rate of 58.7lb/mile, water at the rate of 40.4 gallons/mile and produced 875hp at the drawbar. The corresponding figures for No 1472 were 52.6lb of coal/mile, 38.3 gallons of water/mile and 928hp at the drawbar. Although the Raven 'Pacific' did a better job at maintaining full boiler pressure, these results meant the future of the Raven 'Pacifics' was pretty much decided; another forty Gresley 'Pacific' locomotives were ordered in October 1923. They were considered to be better designed, had more scope for further development and had better valve settings. The selection of the Gresley-designed locomotive 'Pacific' as the L&NER's main express passenger class of locomotive was duly confirmed. They were given the designation of A1 class, with the Raven-designed 'Pacific' locos being given the designation of A2 class.

Only four or five months after the formation of the new company, the ampersand was removed from L&NER, thus making it LNER.

On 27 December 1923, No 1472 entered Doncaster Works for repair because of a fractured centre piston rod. Due to a replacement not being available for some time, the LNER decided to exhibit it at the forthcoming British Empire Exhibition at Wembley. This meant No 1472 had to enter Doncaster Works for preparation to exhibition standards.

Then on 6 February 1924, an official instruction ordered '... that 3000 should be added to all existing GNR locomotive numbers' and so No 1472 became No 4472. It was painted in the new LNER's apple-green livery, with the new coat of arms mounted on the cab sides of the locomotive; brass trim was added to the wheel splashers; the tyres and motion were highly burnished; and finally, the copper and brass fittings were brightly polished. For display at the exhibition it had been coupled to its own original tender No 5223, with the letters LNER surmounting the No 4472 on the tender sides. At this time it was fitted for the first time with nameplates bearing the words 'Flying Scotsman'. No 1472, originally a relatively unknown, mundane locomotive, had been magnificently transformed into No 4472 *Flying Scotsman*. It looked absolutely magnificent – a real show-stopper.

❮ Locomotive No 2054, *Queen Empress* was sent to the Chicago Exhibition in 1893, where it was driven by Crewe driver, Ben Robinson, famous for his record run with *Hardwicke* in the 'Races to the North'. (Ted Talbot collection)

THE
LONDON AND NORTH EASTERN
RAILWAY COMPANY

Three-Cylinder Superheated
4-6-2 PACIFIC TYPE
Express Tender Locomotive

Exhibited at the
BRITISH EMPIRE EXHIBITION
WEMBLEY
1924

❯ The front page from a souvenir book about No 4472 *Flying Scotsman*, when it was exhibited at the 1924 British Empire Exhibition. (Author's collection)

Peter N. Townend, the former pragmatic 'Shed-Master' of King's Cross 'Top Shed' when *Flying Scotsman* was shedded there in the early 1960s, is famous for recommending the fitting of the double Kylchap blast pipe and chimney, plus the accompanying German-style smoke deflectors, to Gresley A3 class locomotives towards the end of Eastern Region steam. Interestingly, his grandfather was responsible for producing a protective cover to wrap up No 4472 *Flying Scotsman* for its journey south to Wembley. So it was that, in its own unique, full-length bespoke hessian 'overcoat' to maintain its pristine condition for the journey to the exhibition site in north-west London, No 4472 *Flying Scotsman* left Doncaster on 2 March 1924 as an almost unknown locomotive at the start of a journey that would make it become 'the most famous steam locomotive in the world'.

'Great Exhibitions', the showing off of a country's greatest achievements, were major events around the world towards the end of the nineteenth century and the beginning of the twentieth century. They had all been triggered by the Great Exhibition of 1851, held in Hyde Park, London. The exhibition was a showcase for the amazing inventions of the time.

Inspired by and wanting to outdo this amazingly successful exhibition of 1851, the Chicago World's Fair of 1893 marked the 400th anniversary of Christopher Columbus' arrival in the New World in 1492. The fair covered an area of more than 600 acres and was attended by more than 27 million people during the six months that it was open. It exceeded previous world fairs in terms of scale and grandeur, and became to the USA what the 'Great Exhibition' was in the UK. The exhibition buildings revived the classical aesthetics and forms of the late seventeenth and eighteenth centuries and were painted white, resulting in the name 'White City' being adopted for the fair site.

For this exhibition, two English locomotives were sent: the GWR received a medal for allowing its 'Iron Duke' class 4-2-2 locomotive *Lord of the Isles* to attend. This class worked on Brunel's 7ft ¼in broad gauge, and was sent even though it was more than a year after Brunel's gauge had been eradicated from the GWR, replaced with Stephenson's standard gauge of 4ft 8½in.

The second locomotive sent was the LNWR's 2-2-2-2 'Great Britain' class three-cylinder compound locomotive No 3435 *Queen Empress*, designed by Francis Webb, along with two carriages. Upon its return to England, it was repainted in lilac and cream livery, renumbered to become No 2054 and was fitted with the royal coat of arms to celebrate the Diamond Jubilee of Queen Victoria in 1897.

For the Chicago Exhibition, the Baltimore & Ohio Railroad Company had constructed an extensive collection of historical railway exhibits, which

› No 4472 *Flying Scotsman* was seen alongside No 4079 *Pendennis Castle* at the second British Empire Exhibition in 1925. Here the two locomotives meet again at *Flying Scotsman*'s new base at Market Overton, under the ownership of Bill McAlpine. (Sir William McAlpine collection)

included about thirty full-size wooden models of the earliest locomotives built in the USA and England, with samples of original tracks.

The Chicago Exhibition in turn inspired other exhibitions to be set up. For example, the area now called White City in West London was named after a building inspired by the previously successful Chicago Exhibition and was used as the site of the Franco-British Exhibition and the Summer Olympics, both of 1908. In 1909, the exhibition site hosted the Imperial International Exhibition, and in 1910 the Japan-British Exhibition. The final two exhibitions to be held at White City were the Latin-British in 1912 and the Anglo-American in 1914. The exhibition building was known as the 'Great White City' due to the white marble cladding used on the exhibition pavilion.

Exhibition fervour continued, with each new display 'borrowing' glory from previous success stories, whilst at the same time trying to outdo the previous exhibitor. Then the First World War came along, putting a lid on any further 'showing off' performances until Britain again wanted to show the world how it should be done!

And so we come to the exhibition that included *Flying Scotsman*. Opened on St George's Day, 23 April 1924, the British Empire Exhibition was the largest exhibition ever held anywhere in the world at that time. Costing £12 million to stage, it was opened by King George V, whose empire contained fifty-eight countries at that time, with only Gambia and Gibraltar choosing not to take part. During the six months that it was open an amazing 27 million visitors toured the site – not bad for a country with a population of about 44 million!

The exhibition's official aim was '… to stimulate trade and to strengthen bonds for all who owe allegiance to the British flag'. The three main buildings of the exhibition were the 'Palace of Industry', 'Palace of Arts' and 'Palace of Engineering'. This last building was the world's largest made from reinforced concrete, a building method that allowed quick construction. The site was linked by several light railways, including the screw-driven 'Never-Stop Railway'. A special railway loop line and station were built to connect the site to London's Marylebone Station.

The 'Palace of Engineering' played host to many representatives from Britain's home railways. The GWR displayed their 'Castle' class 4-6-0 locomotive No 4073 *Caerphilly Castle* – the first locomotive to be constructed at Swindon Works after the First World War. Another first was the Stockton & Darlington Railway's *Locomotion* No 1. A 'Prince of Wales' class 4-6-0 locomotive was specially constructed for the LNWR for the exhibition by the Scottish locomotive manufacturer William Beardmore & Company. London's own Metropolitan Railway was not forgotten and displayed one of their latest 'Inner Circle' underground cars, a first class driving trailer, built in 1923. No 4472 *Flying Scotsman* was added to this roll of honour – going instantly from an unknown quantity with the British public to becoming a great favourite!

After the exhibition ended, *Flying Scotsman* entered Doncaster Works for a heavy repair and was prepared for its second tour of duty at Wembley. This time its original eight-wheeled tender No 5223, which was considered to be too long to fit into the allocated space, was replaced with a Darlington-built K3 class six-wheeled tender, with a capacity of 4,200 gallons. Duly prepared and in position, when the exhibition gates opened for a second season No 4472 *Flying Scotsman* was put on show to the delight of the public once more. The famous engine was accompanied by a group of old and new locomotives, including: GWR's 'Castle' class No 4079 *Pendennis Castle*; the former LNWR's 2-2-2 *Columbine*; and the former Furness Railway's 0-4-0 No 3 *Old Coppernob*, built in 1846.

A few months later, on 17 April 1925, H.N. Gresley announced, at his 'Premium Apprentices and Pupils Annual Dinner' that he would run another set of comparative exchange trials; this time between two types of

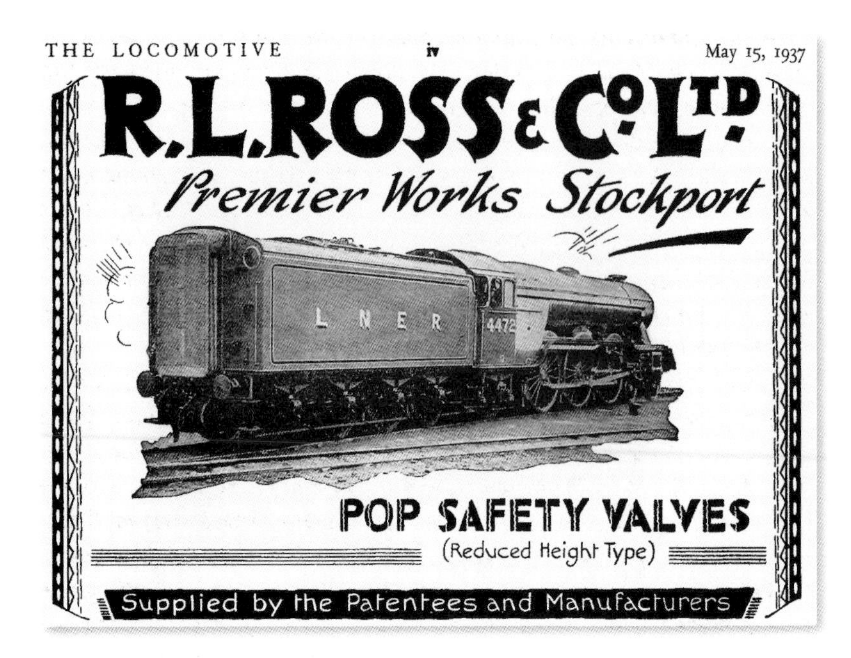

❮ On 5 April 1928, No 4472 *Flying Scotsman* left Doncaster Works after a general repair, during which it had been converted from the generous GNR loading gauge to the LNER loading gauge, allowing it greater route availability. The process involved replacing its original boiler, No 7693, for the new No 7878, with the chimney, cab, dome and safety valves either being replaced or reduced in height. This advert, dated May 1937, from *The Locomotive* magazine, shows where *Flying Scotsman*'s safety valves originated from. (Author's collection)

❯ No 4472 *Flying Scotsman* passes through Doncaster with the heavy 'Harrogate Pullman' dining train. (Author's collection)

locomotive from different regions under the grouping. No 4074 *Caldicot Castle*, built in December 1923 for the GWR, would take part in the trial against LNER's A1 class 'Pacific' No 4474 *Victor Wild*, built in March 1923, with the running taking place between Paddington and Plymouth.

The reason for the trials was that the GWR and LNER, in showing their latest 'state of the art' locomotives, had both claimed that their own locomotive was the more powerful. A set of test runs would settle the matter once and for all. In addition to the Paddington to Plymouth trial detailed above, during April and May 1925 GWR 4-6-0 No 4079 *Pendennis Castle* ran trials against H.N. Gresley's A1 class 'Pacific' No 4475 *Flying Fox* between King's Cross and Doncaster. These trials are discussed earlier in chapter 1. These weeks of locomotive exchanges between the LNER and GWR in 1925 probably proved to have had a greater effect on British locomotive engineering in general than any other comparable event in railway history.

In conclusion, the GWR locomotives emerged as more economical in terms of coal and water consumption, with their superiority being attributed to a higher boiler pressure and, in particular, better valve gear, from which

^ No 4472 *Flying Scotsman* prepares to join its train ready for the inaugural non-stop 'Flying Scotsman' service on 1 May 1928. (Author's collection)

the LNER learned valuable lessons. As a result, a series of modifications were carried out to all the Gresley 'Pacifics' from 1926 onwards, starting with No 4477 *Gay Crusader*, which greatly improved their performance. No 4472 returned to normal main-line service in November 1925 and settled down to the regular work of a Doncaster-based 'Pacific' locomotive, mainly working services between Doncaster and London.

When No 4472 *Flying Scotsman* left Doncaster Works on 5 April 1928 after a general repair, it had been converted from the generous GNR loading gauge to the LNER loading gauge, giving it greater route availability in the north. The process involved replacing its original boiler, No 7693, for the new boiler No 7878, with its chimney, cab, dome and safety valves either being replaced or reduced in height. Short travel valve gear was replaced with long travel valve gear. The coats of arms on the cab sides were removed, its cab-side numbers were restored and it was coupled to corridor tender No 5323, which had the branding LNER on the tender sides.

The economy of the new A1 class locomotives now allowed a change in train working procedures, in that, during the summer timetable of 1928, the principal daytime Anglo-Scottish express, the 10.00 departure, was renamed 'Flying Scotsman' and the service was altered to make it become the longest non-stop passenger service in railway history. The train had been renamed after the locomotive and not vice-versa as is sometimes thought.

On 1 May 1928, No 4472 *Flying Scotsman* worked the inaugural non-stop 'Flying Scotsman' service from London's King's Cross into Edinburgh Waverley, 392.7 miles in 8 hours 3 minutes, with 386 tons tare (unladen weight). King's Cross 'Top Shed' driver Albert Pibworth changed over with Gateshead driver Tom Blades through the newly designed corridor tender at Tollerton, just north of York. The tender suffered from an overheating axle box north of Newcastle and was cooled en route using the slacker pipe. The problem was repaired overnight at Edinburgh.

On the same day, sister locomotive No 2580 *Shotover* departed at 10.00 from Edinburgh Waverley, bound for King's Cross with the inaugural up non-stop 'Flying Scotsman' service, and like No 4472 *Flying Scotsman* became one of five regular locomotives on this service.

The process of steam locomotives picking up water at speed from a trough laid between the running lines to replenish their tenders enabled locomotives such as *Flying Scotsman* to complete non-stop runs from London to Edinburgh and so saved valuable time for the passenger.

In April 1929, No 4472 *Flying Scotsman* co-starred in the first sound feature film to be produced in the UK – *The Flying Scotsman*. The film was directed by Castleton Knight, with Raymond Milland making his first screen appearance as the young fireman in love with the daughter of the driver, Moore Marriott. Marriott, though reasonably acclaimed for this performance, was an actor probably best known as old 'Harbottle' in a number of comedy films that he made with Will Hay and Graham Moffatt, including *Oh, Mr Porter!* and *Ask a Policeman*. Filming with sound came into use halfway through the production, so the comedic thriller has the unusual claim to fame of being part silent era and part 'talkie'. No 4472 *Flying Scotsman* was used by the film company for six weeks of filming, with the film crews using numerous camera positions on the locomotive, tender and rolling stock.

No 4472 *Flying Scotsman* continued to work the non-stop 'Flying Scotsman' service from King's Cross into Edinburgh Waverley through the 1920s and '30s, as well as being used on normal passenger services.

In May 1931, No 4472 *Flying Scotsman* and H.N. Gresley's revolutionary four-cylinder compound 4-6-2-2 locomotive No 10000 moved 'light-engine' (under its own power but with no train) from King's Cross to Nottingham Victoria during a one-day display as part of the Nottingham Civic Exhibition. After appearing there, No 4472 *Flying Scotsman* returned to London and No 10000 returned north. Both locomotives appeared together again in May 1932 at a rolling stock exhibition in Ipswich.

By 30 November 1934, No 4472 *Flying Scotsman* had completed a staggering 653,000 miles since entering service and 44,176 miles since its last general repair. Departing from King's Cross at 09.08, it made a high-speed test run to Leeds with 'six [carriages] on' for a weight of 145 tons tare, travelling a distance of 185.8 miles in 151 minutes 56 seconds. The return trip was achieved in 157 minutes 17 seconds with a load of 205.25 tons tare and included the first authenticated 100mph for steam traction. Driver Sparshatt and Fireman Webster were on the engine for both trips. With the end of the limited speed agreement in 1932, the non-stop 'Flying Scotsman' service time came down to 7 hours 30 minutes, and by 1938 it had dropped to 7 hours 20 minutes.

No 4472 *Flying Scotsman* continued to work the 10.00 King's Cross non-stop 'Flying Scotsman' service into Edinburgh Waverley, but also worked lots of 'bread and butter' duties, including high-speed fish and meat services into King's Cross goods depot originating from Aberdeen. The other prestigious trains worked by *Flying Scotsman* included the 'Harrogate Sunday Pullman', 'Queen of Scots', 'West Riding' and 'Yorkshire Pullman' services until November 1939, after which it entered Doncaster Works for a general repair.

No 4472 *Flying Scotsman*, despite a brief period on station-pilot duties (involving passenger train shunting and assembly) in February 1940,

^ No 4472 *Flying Scotsman* waits at the platform at Newcastle Station for its next turn of duty during the mid-1930s. (W.B. Greenfield, courtesy of the NELPG)

< The fine figure of the front of No 4472 *Flying Scotsman* at Newcastle Station. (W.B. Greenfield, courtesy of the NELPG)

v With such a lot of platform furniture in the foreground, and with a smokey industrial background, No 4472 *Flying Scotsman* is almost lost in this wintry scene at Newcastle in February 1936. (W.B. Greenfield, courtesy of the NELPG)

continued to work services into and out of King's Cross, until on 3 April 1943 the locomotive left Doncaster Works after a general repair, during which it had been repainted in 'war-time' unlined black livery, with the letters NE on the tender.

After the Second World War, the initially straightforward system for the classification of LNER's 'Pacifics' started to break down. In 1945, Edward Thompson rebuilt the first of H.N. Gresley's A1 class locomotives, *Great Northern*. It was initially kept classified as an A1 class locomotive, but on 25 April 1945, the seventeen surviving A1 class 4-6-2s, including No 4472 *Flying Scotsman*, were reclassified to become part of the A10 class. The intention had always been to rebuild the remaining A10s into the new A1s, however this was not done as the rebuild was not successful and instead they were rebuilt to become A3s, with the A10 class becoming extinct in 1948. A brand new class of forty-nine members of Arthur H. Peppercorn's 'Pacific' design of A1 class locomotives were introduced in 1948/9 and, in anticipation of these, *Great Northern* was reclassified as A1/1 class in 1947.

No 4472's duties then became quite varied and, on 24 April 1946, A10 class locomotive No 4472 *Flying Scotsman* was renumbered to become No 502, but in less than two weeks it was renumbered again, becoming No 103. In the midst of all this upheaval, on 5 April 1941 Sir Nigel Gresley sadly died of a heart attack, just two months before he was due to retire.

On 4 January 1947, No 103 *Flying Scotsman* returned to traffic having been rebuilt at Doncaster Works from an A10 class to an A3 class locomotive. In this reclassification it exchanged its boiler, No 7785, for boiler No 8078, which became the sixth boiler that it had carried since it was built. This new boiler had a pressure of 220psi, an improvement of 40psi on the A1 and A10 class boilers, and was fitted with a banjo dome. The tractive effort was increased from 26,926lb to 32,910lb and No 103 was repainted from 'war-time' black into LNER apple-green livery, lined out in black and white.

5

THE BRITISH RAILWAYS YEARS

The magic date for the start of the rebuilding of Britain's railway network after the Second World War was 1 January 1948. By the end of the Second World War, the railway system in Britain was totally worn out, both physically and financially. Clement Attlee's Labour government came to power in July 1945 with a large majority and with a mandate to nationalise the railways and almost every other form of public transport. The Transport Act of 1947 was passed despite substantial opposition and brought virtually all railways, including the London Underground, under the control of the British Transport Commission (BTC).

Although under the auspices of the BTC, it was the Railway Executive that was charged with running the railways. The name 'British Railways' came into immediate use for day-to-day purposes and on 1 January 1948, the nationalised railway system of Britain was born. British Railways, which later traded as British Rail, was the operator of most of the British railway system originating from the nationalisation of the 'Big Four' railway companies in 1948, until privatisation which took place in stages from 1994 to 1997.

Nationalisation saw sweeping changes, with steam traction being eliminated in favour of diesel and electric traction, and passenger services replaced freight as the main source of business. Of course, further change was to follow as a third of the network was deemed to be unsustainable and eventually axed in the aftermath of Dr Beeching's famous report.

Upon nationalisation, the former LNER was sub-divided into three main areas: British Railways Eastern Region, North Eastern Region and part of the newly formed Scottish Region. Prior to nationalisation, LNER locomotives had their depot name painted and even shaded on the bottom right-hand side of the front buffer beam. Afterwards, however, the LMS system of alphanumeric shed-plates attached to a locomotive's smoke box door was adopted. For example, King's Cross went from 'KX', which was often painted as 'Kings +', to become '34A'.

On 15 March 1948, No 103 *Flying Scotsman* left Doncaster Works after a general repair, during which it had exchanged its boiler for one of diagram No 94A, which had a banjo dome steam collector fitted, and became the seventh boiler that it had carried. No 103 was painted in apple-green livery and was renumbered to become No E103, with the legend 'British Railways' painted on the tender.

No E103 *Flying Scotsman* was noted at Newcastle on 5 April 1948 and it continued to work regular up and down services to and from King's Cross, including 'The Yorkshire Pullman'. After receiving a light repair, *Flying Scotsman* left Doncaster Works having been renumbered to become No 60103, retaining its apple-green livery; on 30 December 1948, it still had 'British Railways' on the tender tank sides and resumed its work.

❮ On 15 March 1948, *Flying Scotsman*, now renumbered as E103, left Doncaster Works after a general repair, during which it exchanged its boiler, No 8078, previously fitted to 4-6-2 No 2576 *The White Knight*, for boiler No 9119 fitted with a banjo dome steam collector, previously fitted to No 2505 *Cameronian*. This became the seventh boiler that it had carried. It was painted in apple-green livery, with 'British Railways' on the tender. (Author's collection)

❮ No 60103 *Flying Scotsman* was noted at Leicester Depot on 11 August 1952. It was allocated here from 4 June 1950 until 14 November 1953. (Author's collection)

On 16 December 1949, No 60103 *Flying Scotsman* left Doncaster Works after its latest general repair, during which it exchanged its boiler once more, but this time it had been painted in British Railways blue livery, lined out in black and white and carrying the number No 60103. The early 'cycling lion' crest that was used on locomotives between 1949 and early 1956 replaced the British Railways legend on the tender and in this form it resumed normal duties and also worked an up 'Rugby League Challenge Cup Final' special into King's Cross on 6 May 1950, for the match between Widnes and Warrington, which Warrington won 15–0.

By now *Flying Scotsman*'s shed code allocation had been changed to Leicester Central (38C) and it could be seen working the 'Master Cutler' from Marylebone to Sheffield Victoria, as well as on various other services to Manchester and Nottingham.

Time again for yet another spell in the works, but this time, when No 60103 *Flying Scotsman* left Doncaster Works after a general repair and boiler swap, it was painted in British Railways dark-green livery, not dissimilar in colour to the GWR 'Brunswick green'. Lined out in black and orange, it regularly worked 'The South Yorkshireman' service into Marylebone.

Flying Scotsman was one of the last 'right-hand drive' A3 class locomotives still in active service and was noted in this form on 28 February 1953, departing at 14.10 from Platform A at London Road Station, Manchester (which, after being extensively rebuilt, would be renamed as Piccadilly Station in 1960). No 60103 travelled along the up eastern line towards Ardwick, with an express service into Marylebone. Its H.N. Gresley-designed tender bore the original BR emblem on its sides.

In 1953, *Flying Scotsman*'s depot was changed to Grantham and, in April 1954, *Flying Scotsman* was finally converted to left-hand drive at Doncaster Works.

Notable services worked by *Flying Scotsman* in the 1950s were: 'Heart of Midlothian', 'Norseman', 'Northumbrian', 'Queen of Scots', 'Scarborough Flyer', 'Talisman', 'Tees-Tyne Pullman', 'Tyne Commission

Quay', 'Tynesider', 'White Rose', 'Yorkshire Pullman' and, of course, the 'Flying Scotsman' – you name it and *Flying Scotsman* probably worked it.

After all this hard work, *Flying Scotsman* arrived at Doncaster Works again on 10 December 1959 for another boiler change. It left several weeks later, fitted with a Kylchap double blast pipe and a double chimney; all the work was accomplished at a cost of £153. After many more 'Talisman' and 'Yorkshire Pullman' workings, it started working the 'Aberdonian' and 'Car-Sleeper' services until January 1960, when on the 27th No 60103 *Flying Scotsman*, with '11-on' (pulling eleven carriages) and some 400 tons behind the tender, including the usual heavy and comfortable coaches, plus restaurant cars, worked the 13.00 King's Cross 'Heart of Midlothian' service for Edinburgh Waverley. A slightly late start was made, but with good uphill work and expert handling by a King's Cross top-link driver, the train didn't exceed 80–81mph and stopped at Peterborough before time, in 74 minutes for the 76½ miles , where No 60103 was taken off the train. Now behind A2 class 4-6-2 No 60500 *Edward Thompson*, the train departed for York, which was the next stop, with *Edward Thompson* providing the power as far as Newcastle.

Flying Scotsman continued to earn its keep and entered Doncaster Works once more for another boiler change, and this time it emerged with the fourteenth boiler that it had carried.

On 10 April 1961, No 60103 *Flying Scotsman* worked the 14.00 King's Cross 'Tees-Thames' service, on diagram No 74, but its next interesting working was on 6 May, when it worked the Gainsborough Model Railway Society's first ever special, 'The Lake District Railtour'. The train, which started at Lincoln, travelled via Doncaster, York, Gateshead, Newcastle and Hexham to Carlisle and return. The GMRS would subsequently prove to get even more involved in the subsequent chapters detailing the history of *Flying Scotsman*.

Since the fitting of the double Kylchap blast pipe and double chimney, experience with smoke drifting over the cab of A3 class locomotives was becoming a problem. Peter N. Townend, the pragmatic 'Shed-Master' of King's Cross, recommended to the 'powers that be' that German-style trough smoke deflectors should be fitted to ease the problem. Several locomotives were so fitted and the experiment was deemed to be a total success. On 16 December 1961, No 60103 *Flying Scotsman* left Doncaster Works after a casual light repair, during which it had been fitted with German-style trough smoke deflectors. Essentially they were a variation of one of two designs of the *Windleitbleche* (literally, 'smoke deflector') originating from the *Deutsche Reichsbahn-Gesellschaft* – the German national railway company between the two world wars. No 60103 was

^ On 11 May 1952, No 60103 *Flying Scotsman* was noted at Sheffield Darnall Depot (41A), after which it then worked an up Manchester service into Marylebone from Sheffield. (Author's collection)

then cleaned and received a coat of BR dark-green paint, after which it returned to its home depot of King's Cross (34A).

Although No 60103 continued its daily duty of working many passenger trains, names of which have been listed previously, the service life of *Flying Scotsman* working for British Railways was coming to an end, with the wholesale introduction of diesel-electric powered locomotives.

When the last day of service of *Flying Scotsman* finally arrived, something special happened to this famous locomotive: instead of going to the breaker's yard to be dismantled for scrap, it was sold to a private buyer, who gained permission to run it on British Railways tracks and, in the process, saved this icon for future generations to enjoy.

On 14 January 1963, in the middle of a 'big freeze', the preparation of No 60103 *Flying Scotsman* at King's Cross 'Top Shed' was captured by TV and press cameras, including a number from the USA and Canada. Again and again Alan Pegler retold the story of how he came to '… save the "lady of his dreams" from a fate worse than death – ignominious dispatching to the scrap-heap'. With scenes reminiscent of a royal occasion, Alan Pegler and the train crew wore carnations as the King's Cross Station Master,

On 3 April 1955, No 60103 *Flying Scotsman* worked a 'Westminster Bank Railway Society Special', hauling six vintage GNR coaches, a cafeteria car and the 'Coronation' Beaver Tail observation car. The route was King's Cross–Hitchin–Grantham–York, taking 3½ hours. The return journey departed from York at 18.05, but due to engineering works arrived into King's Cross at 21.28. (Author's collection)

On 14 March 1952, No 60103 *Flying Scotsman* left Doncaster Works after another general repair. It had been painted in British Railways' green livery, often referred to as 'Brunswick green', but more correctly as 'middle chrome green' and lined out in black and orange. (Sir William McAlpine collection)

No 60103 *Flying Scotsman* is seen at Doncaster Carr locomotive depot (36A) during 1958. (Author's collection)

No 60103 *Flying Scotsman*, is seen at King's Cross Depot (34A) on 30 May 1959, after it worked the 11.20 down King's Cross service to Scarborough. It had by this time been fitted with a double chimney. (Author's collection)

^ No 60103 *Flying Scotsman* is seen at Woolmer Green hauling nine Pullman cars, working the down 'Yorkshire Pullman' service on 25 July 1959. (A.E. Buckley)

^ No 60103 *Flying Scotsman* is seen passing through Hornsey with a down King's Cross Pullman service during the autumn of 1960. (Author's collection)

^ On 21 February 1960, No 60103 *Flying Scotsman* was again noted at King's Cross Depot (34A). (Author's collection)

^ No 60103 *Flying Scotsman* is seen in 1961, working an express passenger service past Finsbury Park signal box. (Author's collection)

^ No 60103 *Flying Scotsman* is seen on shed at York on 5 September 1954, painted in British Railways green livery, lined out in black and orange. (Frank Hornby)

^ Standing at King's Cross Station on 14 January 1963, No 60103 *Flying Scotsman* awaits its last official passenger train service under BR ownership – the 13.15 from King's Cross to Leeds. (D. Trevor Rowe)

› On 21 April 1960, No 60103 *Flying Scotsman* worked an 'Ian Allan Locospotter's Special' between Marylebone and Doncaster, via Rugby Central and Bagthorpe Junction, Nottingham. At Doncaster, 'loco-spotters' were given a tour of the works, before returning to Marylebone. Here the train is seen at Leicester. (Author's collection)

wearing his top hat, read out messages of congratulations to Alan Pegler's wife, Pauline. No 60103 *Flying Scotsman*, with '11-on', then worked its last official passenger train service under BR ownership, the 13.15 from King's Cross to Leeds. The route was: King's Cross–Peterborough–Doncaster (arriving at 15.47, where *Flying Scotsman*'s journey would end). The service then went on to Wakefield and Leeds.

Along most of the route to Peterborough, railwaymen and enthusiasts were out to give a wave and a cheer, and a maximum speed of 90mph was obtained while passengers were having lunch. After a quick stop at Peterborough, No 60103 departed for the next stop at Doncaster and arrived 4 minutes early at Retford. Three miles north of Retford at Barnby Moor level crossing, a party of Alan Pegler's friends had turned out to see the latest Pegler acquisition, which called for a lot of waving by Alan and Pauline from the train and a long blast from the whistle. Symbolically, the sun was now setting in a frosty sky as *Flying Scotsman* approached Doncaster, where it came to a stand exactly 6 minutes early. Then, after its 2-hour 32-minute journey to Doncaster, No 60103 *Flying Scotsman* was taken off the service and moved into Doncaster Works, having travelled some 2,076,000 miles in public service since originally being completed in 1923.

So as the sun set on another day, another chapter in the history of this famous locomotive came to an end.

But was the best yet to come?

6

THE ALAN PEGLER YEARS

Penny Pegler, daughter of Alan Pegler, describes how *Flying Scotsman* came into her life to change it forever:

It was a snowy January evening in 1963 and I was 9 years old. My father came upstairs as always, to read me my bedside story, but when he popped his head around the door there was a twinkle in his eye. He often had a twinkle, but this seemed to be a special twinkle and was accompanied by a mischievous grin. He sat down on my bed, and whispered 'Today I bought a steam engine! She is called "*Flying Scotsman*" and I will have her painted apple green.'

He went on to describe how she was destined to be destroyed and that there were no identical engines anywhere in the world. He was stepping in to save her. This all seemed perfectly natural to me, as Papa was amongst the saviours of my friend 'Prince' and the Ffestiniog Railway, so I snuggled down to sleep imagining what adventures lay ahead of us, but wondering slightly where we would keep our bright green steam locomotive.

The problem was soon resolved: we were to use a shed in Doncaster. Saturday mornings would see us gathering up picnic baskets and cotton rags and driving off to 'our shed', where we would clean and polish. At the end of the day we would say goodbye to 4472, as she had become, pile into the car and drive home with grimy clothes and happy faces.

By now she had been nicknamed 'Scotty' and was part of the family. Little did we realise that this wonderful locomotive would change the course of our lives.

• • •

On 14 January 1963, Alan Pegler travelled up to King's Cross from Retford on the 'Master Cutler' Pullman service, arriving in London soon after 10.00. Transport had been laid on to take him out to 'Top Shed', where the preparation of No 60103 *Flying Scotsman* in freezing temperatures was captured by cameras of the assembled media. Alan's wife, Pauline, told the *Daily Sketch*: 'Trains are not just a hobby for my husband. You could call them a professional life interest.'

As described in the previous chapter, No 60103 was taken off its final train at Doncaster Station and moved off into Doncaster Works, ending forty years of public service since its original completion in 1923. Upon entering the works, *Flying Scotsman* had entered the world of preservation.

During *Flying Scotsman*'s time inside Doncaster Works, Alan had arranged for it to be restored as closely as possible to its former LNER

HEADS OF AGREEMENT

between the British Railways Board and Alan Francis Pegler, Esq., relating to the sale to him of the 'Flying Scotsman' railway locomotive.

1. In these Heads of Agreement:
'the Board' means the British Railways Board and (where the context so admits) its legal successors;
'the Purchaser' means Mr. Alan Francis Pegler and (where the context so admits) his successors in title;
'the locomotive' means the Board's railway locomotive No. 60103 known as 'Flying Scotsman' and includes the tender thereof.

2. The Board will sell and the Purchaser will purchase free from incumbrances the locomotive at the price of Three thousand pounds (£3,000.0.0). The property in the locomotive will pass on delivery which will take place at noon on 16th April, 1963 at King's Cross Motive Power Depot, London.

3. Prior to delivery or as soon thereafter as may be the Board will at their own cost:
(a) remove the existing smoke deflector plates from the locomotive;
(b) replace the existing chimney and blast pipe by a single chimney and blast pipe similar to those with which the locomotive was formerly fitted;
(c) replace the existing tender by a former L.N.E.R. corridor tender;
(d) repaint the locomotive and the tender in L.N.E.R. express passenger locomotive colours;
(e) restore to the locomotive its former number, viz: 4472;
(f) ensure that the locomotive is in running order and carry out two satisfactory trial runs.

4. The locomotive when not in use upon the railway will be stabled in the Old Engine Weighhouse at Doncaster Loco. Works at an annual rent of Sixty five pounds (£65.0.0) or at such other place as the Board may from time to time determine. The Purchaser shall be liable for any rates from time to time payable in respect of such premises and in addition will bear the cost of keeping the same in good and substantial repair and condition to the satisfaction of the Board.

5. If the Purchaser, or any duly authorised representative of his, wishes to visit the locomotive while stabled or to carry out any inspection thereof he shall make the necessary arrangements with the Board's Works Manager, Doncaster. Apart from personnel of the Board who may in connection with the provisions hereof require so to do neither the Board nor the Purchaser shall without the consent of the other permit any other persons (and in particular any member of the public) to visit the locomotive while stabled.

6. The Purchaser shall from time to time and at his own cost make all necessary arrangements with the Board's Traffic Manager, Doncaster for the proper maintenance of the locomotive and all necessary periodical inspections thereof so as to ensure that the locomotive is kept in good repair and condition and that all legal requirements are complied with. However, as a matter of practical convenience whenever the locomotive is used for the purposes of the business of the Board in accordance with the provisions for hiring the locomotive contained in Clause 8 the Board will at a reasonable cost to the Purchaser properly maintain and repair the locomotive.

7. The locomotive will be stabled at the Purchaser's risk in all respects and it will be for the Purchaser, if he so desires, to effect any insurance with regard to the locomotive. In the event of a hiring out of the locomotive for the purposes of the business of the Board the Board shall keep the Purchaser fully indemnified against all actions claims and demands taken or made by any person and against any loss incurred by the Purchaser arising out of or in connection with any accident or event whatsoever occurring during the period of hiring so however that this indemnity shall not be construed as extending to the making good of any fair wear and tear.

8. Subject to Clause 13(a) in the event of:
(a) the Purchaser wishing the locomotive to be run over any part of the Board's railway system; or
(b) the Board wishing to use the locomotive;
the Purchaser or the Board (as the case may be) shall give reasonable notice to the other of their proposal, together with all relevant information. If any such proposal is agreed between the parties the transaction (in so far as it relates to the user of the locomotive) shall take the form of a hiring of the locomotive to the Board from the Purchaser upon the terms and conditions then agreed between them. Each party shall have complete discretion either to refuse any proposal put forward by the other or only to agree thereto subject to such conditions as the Purchaser or the Board (as the case may be) think fit.

9. Subject to the provisions of Clauses 6, 8 and 13(a) hereof and to the provisions of this clause and except in the case of emergency no person may maintain, repair, drive or fire the locomotive, or ride thereon without the consent of the Board and the Purchaser. In particular the Board will not issue any footplate pass without first consulting the Purchaser. However nothing in these Heads of Agreement shall prevent the Board from authorising any of their personnel, in the course of their normal duties, to ride upon the footplate of the locomotive in connection with the proper discharge by the Board of any obligations arising under or by virtue of this Agreement.

10. The Board shall not at any time hereafter use or permit the use of the name 'Flying Scotsman' on or in connection with any steam locomotive owned or controlled by it in the United Kingdom but nothing herein contained shall prevent the Board from using such name to describe any train service run by it.

11. The Board hereby granted to the Purchaser full but non-exclusive royalty free licence to use the locomotive under any patents and registered designs covering the locomotive or any part thereof.

12. This Agreement other than Clauses 2 and 3 thereof may without prejudice to the accrued rights of either party be determined:
(a) by either party giving to the other not less than twelve months' notice in writing expiring in or at any time after the Sixteenth day of April One thousand nine hundred and sixty six; or
(b) forthwith by the Board in the event of any sum due hereunder by the Purchaser to the Board remaining unpaid for more than six months.

13. In the event of this Agreement being determined:
(a) under Clause 12(a) hereof and by the time the notice has expired no arrangements have been made for the stabling of the locomotive elsewhere (for which purposes the Board shall provide all reasonable assistance and shall grant to the Purchaser the right to run the locomotive over the Board's railway system to such extent as may be necessary); or
(b) under Clause 12(b) hereof;
the Board may forthwith re-possess the locomotive and sell the same provided that the Board shall account to the Purchaser for any proceeds of sale less any sums properly due from the Purchaser to the Board, including the cost of such re-possession and sale.

DATED this Sixteenth day of April One thousand nine hundred and sixty three.

Signed
by the Purchaser ....

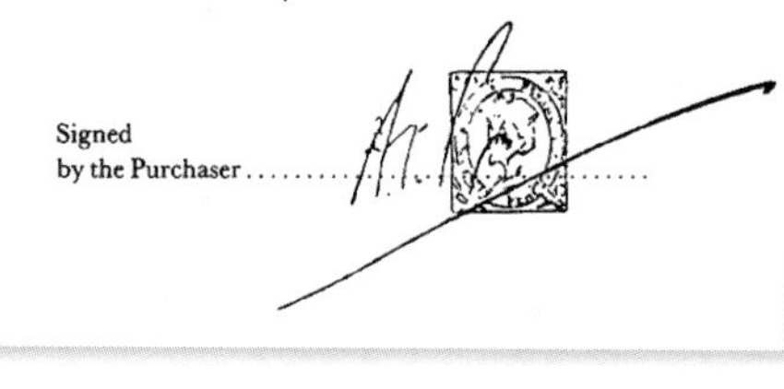

❮ The agreement reached between the British Railways Board and Alan Pegler, relating to the sale of *Flying Scotsman*. (David Ward collection)

condition: the German-style smoke deflectors were removed, the double chimney was replaced by a single chimney and the tender was replaced with one of the corridor type, similar to that with which it had operated between 1928 and 1936 in the years of the non-stop runs between King's Cross and Edinburgh Waverley. *Flying Scotsman* was also repainted into its former LNER livery of apple green and had its former number 4472 reinstated.

Part of the contract to buy *Flying Scotsman* allowed Alan Pegler to run it on BR tracks, with BR maintaining the locomotive in working order. With this in place, No 4472 *Flying Scotsman* was therefore able to work a considerable number of private charter 'rail tours' all around the country.

During this period, the watering facilities for steam locomotives were fast disappearing, so to obviate this short-coming, Alan Pegler purchased a second corridor tender in September 1966 and, retaining its side corridor, had it adapted as an auxiliary water tender. When in use, it was coupled behind the normal tender and was coupled in such a manner so that the water would flow into *Flying Scotsman*'s normal tender on the move. Purchasing and modifying the second tender to carry water set Alan back £6,000, a very large amount considering *Flying Scotsman* itself had only cost £3,500.

In 1968, the year that steam traction officially ended on British Railways, *Flying Scotsman* ran a non-stop run from London's King's Cross to Edinburgh Waverley, to commemorate the fortieth anniversary of the first non-stop 'Flying Scotsman' service.

As mentioned, as part of the deal to buy *Flying Scotsman*, Alan Pegler had signed a contract permitting him to run his locomotive on BR tracks.

❮ *Vera*, one of three former 'Brighton Belle' vehicles in the Venice Simplon-Orient-Express British Pullman fleet, is seen at Platform 2, Victoria Station, waiting for another departure. *Flying Scotsman* worked this train many times at the start of the twenty-first century. (Author)

❮ No 4468 *Mallard*, the world speed record holder for steam traction, is another contender for the title of 'the most famous steam locomotive in the world'. (Author's collection)

❮ No 4472 *Flying Scotsman* and its exhibition train wend their way on the early stages of their North American tour in 1969. (The Southern Museum of Civil War and Locomotive History, Kennesaw, Georgia collection)

❮ The LNER coat of arms was carried on the cab sides of *Flying Scotsman* on its North American tour. The author and Sir William McAlpine are each proud owners of these crests, from the fireman's side and driver's side respectively. (Jack Neville)

⌄ On 18 March 1972, No 4472 *Flying Scotsman* was at Pier 41 on the Embarcadero in San Francisco, with part of the exhibition train. Note *Britannia* behind the second tender, a steam vehicle which had been driven halfway around the world by its owner from England, and the *Balclutha*, built in 1886 and the last steel-hulled fully rigged sailing ship left in the San Francisco Bay area. (Jack Neville)

^ In a well-advertised move, No 4472 *Flying Scotsman* was ferried back to Oakland from San Francisco on the ferry *Las Plumas*. No 4472 and its train are seen being worked behind Western Pacific U-23B class diesel-electric locomotive No 2260, at the junction of 45th and Leandro streets in West Oakland. (Jack Neville)

› A British Railways flyer, advertising the 'Cumbrian Coast Express'. As advertised, *Flying Scotsman* would work the train in one direction and *Sir Nigel Gresley* would work in the opposite direction between Carnforth – where the two locomotives were based – and Sellafield, the site of the nuclear reprocessing plant. (David Ward collection)

⌄ Resplendent in its shiny apple-green livery, No 4472 *Flying Scotsman* and its support coach simmer gently in the York sunshine, waiting for their next turn of duty during the McAlpine era. (Sir William McAlpine collection)

› A flyer promoting the meeting of *Flying Scotsman* and its old nemesis *Pendennis Castle* in Perth, Western Australia, during 1989. (Sir William McAlpine collection)

^ Newly restored as No 4472 *Flying Scotsman* is seen outside the National Railway Museum at York, being prepared for its return journey to London on 11 July 1999. Note that the cab and tender have still not been lined out. (Grahame Plater)

^ No 4472 *Flying Scotsman* is seen taking on coal, Great Western Railway-style, at the Didcot Railway Centre, during Tony Marchington's ownership. (Grahame Plater)

^ Here is a general view of the frames of *Flying Scotsman* during the Tony Marchington-era overhaul, which was carried out at Southall during the late 1990s. (Grahame Plater)

^ This picture shows the attention to detail that was being carried out on the frames of *Flying Scotsman* during the spring of 2012, under NRM ownership, when it was at the works of Riley & Son, Bury. (Author)

❮ By the end of March 2012, work was completed on the design, manufacture and riveting of the new mid-stretcher for *Flying Scotsman*'s frames at the Riley & Son works. (Author)

^ Another contender for the title of 'the most famous steam locomotive in the world' may be 'Pacific' locomotive *Tornado*. Numbered 60163, it is seen in BR blue livery as it attacks Grosvenor Bank on 29 November 2012, working 'The Cathedrals Express' charter from Victoria Station, London, to Bristol and return. *Tornado*'s design is a derivative of that of *Flying Scotsman*. (Author)

› Built at Springburn Works, Glasgow, in 1886, Caledonian Railway 4-4-0 No 123 took only sixty-six days to construct and was an example of cutting-edge locomotive technology. Its greatest claim to fame was its participation in the celebrated first 'Race to the North' of 1888, in which east coast and west coast railway companies competed to get passengers to Edinburgh in the fastest time. (Author's collection)

❮ The GNR's small-boilered C2 class locomotive was the first 4-4-2, or 'Atlantic'-type, locomotive to be introduced to Great Britain. Designed by Henry Ivatt in 1897, a total of twenty-two were built between 1898 and 1903 at Doncaster Works. The locomotives were known as 'Klondikes', after the 1897 Klondike gold rush. (Author's collection)

❮ In December 1902 the first of H.A. Ivatt's large 'Atlantic' locomotives, No 251, entered service in Great Britain. The C1 class had been developed as an enlarged version of what became the LNER's C2 class for use on the fastest and heaviest express passenger trains on the GNR. These large-boilered 'Atlantics' became the first locomotives to be built with a wide firebox, effectively making them the beginning of the East Coast Main Line's 'big engine' policy. (Author's collection)

❮ By 1888 the fastest journey time between London and Edinburgh had been reduced to 7 hours 27 minutes. But, with the opening of the Forth Railway Bridge in 1890, there was renewed competition to get even further north to Aberdeen. During the night of 22 August 1895, the LNWR and the Caledonian Railway ran an especially light train from London to Edinburgh in 512 minutes, beating their east coast competitor's time by 8 minutes, with *Hardwicke* hauling the train the 141 miles from Crewe to Carlisle in 126 minutes, including the 900ft Shap Summit climb. (Author's collection)

› Two locomotives, a 2-2-2 and a 4-2-2 built for the GNR, are seen double-heading the forerunner of the 'Flying Scotsman' train on the East Coast route. Not only would *Flying Scotsman* ultimately replace these two locomotives, but it would complete the journey from London to King's Cross non-stop and in a quicker time. (Author's collection)

› The NER's rival to the GNR's 'Pacifics', No 2400 *City Of Newcastle* was designed by the NER's Chief Mechanical Engineer Sir Vincent Raven and was built at their Darlington North Road Works in 1922. (Author's collection)

› A contemporary postcard of No 4472 *Flying Scotsman* shows how it looked when it was exhibited at the British Empire Exhibition, Wembley in 1924. (Author's collection)

❮ *The Flying Scotsman* is a romantic-thriller-comedy which is famous for being the first 'talkie' film to be produced in the UK. It is notable for Ray Milland's first screen role, as well as for the daring stunt work done on the train itself. Here, Ray Milland looks at No 4472 *Flying Scotsman* during a trip celebrating the locomotive's sixtieth birthday. (Sir William McAlpine collection)

❮ No 60103 *Flying Scotsman* at King's Cross in 1962. During a casual light repair in December 1961, No 60103 was fitted with German-style trough smoke deflectors, under the recommendation of Peter N. Townend, who was the 'shed master' at King's Cross Depot. The fitting of a double chimney caused a softer exhaust beat, allowing smoke to drift down over the cab, obscuring the driver's vision, and so deflectors were fitted to prevent this problem. Although they had dubious aesthetic appeal to some, they were a complete success. (Geoff Rixon)

❮ No 60103 *Flying Scotsman* in its new guise is seen working a train with twelve coaches during 1962. (Author's collection)

^ A classic view of a classic locomotive. No 4472 *Flying Scotsman* on the Forth Railway Bridge during one of its many visits while owned by Alan Pegler. (Author's collection)

< Sparkling in the sunshine, the former LNER No 4472 *Flying Scotsman* is prepared on what was former GWR territory for yet another special charter during the Alan Pegler era. (Author's collection)

❮ *Flying Scotsman* with two tenders on the Anglo-Norse Society's Norfolk Enterprise rail tour on 12 May 1968. The Norwegian flag is among those at the front of the locomotive. (Author's collection)

⌄ No 4472 *Flying Scotsman* waits at King's Cross Station ready to depart with 'The Norfolkman', bound for Norwich, on 20 May 1967. (Mark Dixon collection)

❮ A close-up at Carnforth of the classic American 'pilot', known as a 'cowcatcher' in Britain. This was fitted to *Flying Scotsman* for its North American tour, and is now in the author's possession. (Author)

⌄ No 4472 *Flying Scotsman* in full flight with its exhibition train during the early stages of the North American tour. (Paul Dowie collection)

^ *Flying Scotsman* arrived in San Francisco in March 1972. This shot was taken on that first weekend at Maritime Park, at the Hyde and Beach Street terminus of the Powell Street cable car line. San Francisco Bay is visible in the background. (Jack Neville)

❮ This sign was more reminiscent of an advert proclaiming that 'the circus is in town' – which in a way it was! The observation car, SC 281, formerly Pullman car No 14, waits for its next turn of duty carrying tourists along the Embarcadero in San Francisco. (Edward Saalig)

⌄ During 1971 and 1972, No 4472 *Flying Scotsman*, its administration car, Pullman Car *Lydia* and observation car worked trains at 10mph along the Embarcadero at weekends from 10.00 until 18.00. Here they are seen at the west end waiting for an obstacle to be moved out of their way so the train can continue. The California Shell Fish Company looks exactly the same today. (Jack Neville)

❮ During March 1963, *Flying Scotsman* was being repainted in its former LNER apple-green livery. The Doncaster paint shop foreman, Mr Parsons, arranged for the nameplates to be painted red without consulting Alan Pegler. According to Mr Parsons: '... it was Nigel Gresley himself who had suggested this scheme for special locomotives, but the war and Nigel Gresley's death meant that no locomotives had actually ever had a red nameplate.' Mr Parsons thought that '... this would be the only opportunity that this idea would be put into practice'. (Edward Saalig)

⌄ No 4472 *Flying Scotsman* and her train were placed in storage at Sharpe Army Depot in Lathrop on 12 August 1972, where the ensemble is seen having arrived mid-afternoon. (Jack Neville)

› Universal Studios Hollywood is one of the oldest and most famous Hollywood movie studios still in use. It was here from 1977–97 that several coaches from *Flying Scotsman*'s North American tour exhibition train were used as dining coaches in the Victoria Station Restaurant. (Author)

› This is an interior view of the Victoria Station Restaurant at Universal Studios, California, not long after its original opening in 1977. On the left can be seen the administration car and on the right is one of the Gresley-designed Full-Brake vans. They were both painted in unlined green and cream livery. (Paul Dowie collection)

^ Reception car No E 104 E, seen at the Victoria Station Restaurant, was built in 1948 at York to an Edward Thompson design. No E 104 E's primary use in the North American tour was as a means for visitors to enter the exhibition train. It had, however, been fitted out with a branch of Lloyds Bank and was also used for the storage of literature. It was still being used as a dining car as late as 1997, when the restaurant was refurbished and all traces of the restaurant complex disappeared completely. (Author)

❯ On 13 February 1973, No 4472 *Flying Scotsman* arrived back in England at Liverpool aboard the *California Star*, after its North American adventures. Here is a first day cover commemorating the event. (Author's collection)

❯ No 4472 *Flying Scotsman* stood outside Derby Works from 19 February to 14 July 1973, including a period mantled in snow and awaiting attention. No 4472 was then moved inside Derby Works where it received an inspection and a full repaint in LNER livery, when the nameplates reverted to black, as did the cylinder covers. The damage caused by the Californian sun to *Flying Scotsman*'s paintwork can be clearly seen. (Sir William McAlpine collection)

❮ No 4472 *Flying Scotsman* is seen at Market Overton during its short time there in 1974, having been moved there by its owner Bill McAlpine. (Sir William McAlpine collection)

↳ No 4472 *Flying Scotsman* is seen on the Carnforth turntable during the 1980s. Towards the upper left corner can be seen the spare boiler for *Flying Scotsman*, waiting to be used in a future overhaul. (David Ward collection)

⌄ A vista enjoyed by only a few. This is the view from the fireman's side of *Flying Scotsman*'s cab, as it was working an enthusiasts' special during the 1980s. (David Ward collection)

❮ Two of Herbert Nigel Gresley's masterpieces in one picture. On the right is A3 class 'Pacific' No 4472 *Flying Scotsman*; on the left is A4 class 'Pacific' No 4498 *Sir Nigel Gresley*, the hundredth 'Pacific' locomotive to be built under his design. They are at Carnforth, their base during the 1970s and 1980s for working various enthusiasts' specials. (David Ward collection)

❮ No 4472 *Flying Scotsman* is mobbed as it stops at Dapto, during its successful tour of Australia as part of the celebrations of that nation's bicentenary in 1988. (Author's collection)

↖ As can be seen, people were willing to go the extra mile to see No 4472 *Flying Scotsman* during its successful Australian tour in the late 1980s. (Sir William McAlpine collection)

↑ Steaming out of Sydney's Central Station under cloudy skies and a cool breeze, No 4472 *Flying Scotsman* piloted 38 class 'Pacific' No 3801 on a run of nearly 687 miles bound for Brisbane, the capital of Queensland, the Sunshine State. The two green locomotives were in charge of a twelve-coach train, with 200 passengers on board. The route was Sydney–Maitland–Musswellbrook–Maitland. (David Ward collection)

‹ Here is the headboard that was carried on the front of *Flying Scotsman* during its amazing trip to Australia. It is seen when it was displayed on the wall of Southall Depot's workshop, but has since been moved to the workshops of Sir William McAlpine at his Fawley Hill Railway Centre. (Author)

^ On 19 August 1989, just south of Alice Springs, a brief parallel run took place with No 4472 *Flying Scotsman* and a 4-8-2 steam locomotive No 124, of the recently restored 3ft 6in gauge 'Old Ghan' line that originally ran from Port Augusta. (Sir William McAlpine collection)

> On 14 December 1989, No 4472 Flying Scotsman was unloaded from *La Perouse* at Tilbury, after its journey from Sydney, completing the first known circumnavigation of the globe by a steam locomotive. 'Moveright International' transported No 4472 *Flying Scotsman* from Tilbury Docks to Seabrook Sidings, after which *Flying Scotsman* was 'tripped back' to its home depot of Southall. (Sir William McAlpine collection)

❮ No 60103 *Flying Scotsman* runs around its train at Swanage, during a visit there in 1994. (Author)

❮ Another view of No 60103 *Flying Scotsman* at Swanage, showing the cab to good effect. (Author)

❮ After its overhaul, as No 60103 *Flying Scotsman* wasn't certificated to run on the public railway network, it was moved by road to the Paignton & Dartmouth Railway and recommenced its tour of private railways, working various passenger duties. On 14 September 1993 it ended its tour at Paignton, having completed 2,730 miles in service. *Flying Scotsman* is seen crossing Churston Viaduct with a passenger service. (D. Trevor Rowe)

^ In June 1995, the dismantling of No 60103 *Flying Scotsman* was authorised in preparation for its next major overhaul. Having had the external plate work removed, it is seen here having the boiler insulation, consisting of rock wool, removed by some of *Flying Scotsman*'s volunteer team. (Fred Stenle)

An idyllic view of No 60103 *Flying Scotsman* on the Nene Valley Railway in the 1990s. (D. Trevor Rowe)

This view of the stripping down of *Flying Scotsman* at Southall, during the Marchington era, shows that the process is in full swing, with the boiler, cab, motion, two outside cylinders and two sets of driving wheels having been removed – but there was still more to do. (Fred Stenle)

❮ In this view, the buffer beam has been stripped back to base metal – with the exception of the 'A3', *Flying Scotsman*'s class designation. A closer look will show the rivet heads holding the buffer beam to the various sections of the frames. (Author)

❯ In this view, *Flying Scotsman*'s refurbishment has advanced so much that the 'bottom end' is all but complete. Next job would be the fitting of the boiler. (Fred Stenle)

❮ During June 1999, *Flying Scotsman* went through a period of commissioning tests, consisting of static steaming and short runs up and down Southall Depot yard without leaving the depot. An estimated distance of 10 miles was achieved during this period, with the locomotive in steam for 80 hours or so. *Flying Scotsman* is seen being prepared for another test. (Fred Stenle)

^ No 4472 *Flying Scotsman* at its home depot of Southall on 3 July 1999, with the paint hardly dry and before lining out. (Derek Crunkhorn)

^ No 4472 *Flying Scotsman* works the 'Inaugural Scotsman' special charter, from King's Cross to York, through Finsbury Park Station on 4 July 1999. (Author)

❮ The author is seen filming and interviewing Roland Kennington, the chief engineer of *Flying Scotsman*, in the yard at the NRM, York, shortly after *Flying Scotsman* had successfully completed the 'Inaugural Scotsman' run. (Grahame Plater)

❮ In magnificent condition, both mechanically and externally, No 4472 *Flying Scotsman* is seen working one of its first enthusiasts' special workings after the most expensive and extensive overhauls ever carried out on a steam locomotive in private ownership. (Author's collection)

^ The unsung heroes of main-line preserved steam traction – the support crew – seen at Salisbury after *Flying Scotsman* had worked the 'Sarum Scotsman' charter from Paddington to Salisbury in August 1999. (Author)

A sneak peek inside *Flying Scotsman*'s smoke box reveals part of the Kylchap blast pipe surrounded by the spark arrester. (Grahame Plater)

As No 4472 *Flying Scotsman* waits to depart from Slough during 1999, there is somewhat more than enough steam to spare. This phenomenon is occurring more and more in the preservation era and is perhaps due to the inexperience of the crews. (Grahame Plater)

^ On 5 November 2000, No 4472 *Flying Scotsman* was due to work a charter from London to Stratford-upon-Avon, but the trip was cancelled. So, entering into the spirit of the day, a support crew member had set off a Catherine wheel on *Flying Scotsman*'s smoke box door. (Grahame Plater)

⌃ No 4472 *Flying Scotsman* at Platform 1, King's Cross Station, and now fitted with German-style smoke deflectors, which, although not historically correct alongside the apple-green livery, were certainly necessary to improve the driver's vision. (Author's collection)

❮ No 4472 *Flying Scotsman* stands at Stewarts Lane Depot, with its smoke box door open, revealing a sight not normally seen by the general public – the ashes and dust left over from the burned coal of a previous run. This debris removal is part of the regular preparations before the next outing. (Fred Stenle)

› 2003 marked the 150th anniversary of the opening of the GNR's works at Doncaster, known locally and throughout the railway industry as 'The Plant'. *Flying Scotsman* is seen at Doncaster Works, outside its 'birthplace', the day before railway enthusiasts visited the open weekend celebrating this milestone. (Author)

^ Some people consider George Stephenson's *Rocket* the most famous steam locomotive in the world. Seen outside the Museum of British Transport, Clapham, is a replica of the famous engine. When the museum closed in 1973, this replica was used to construct the NRM's working replica of 1979 and is still working, visiting various museums around the country, and indeed the world. (Author's collection)

^ The Beeching Report, published on 27 March 1963, recommended that British Railways should stop running museums and so a campaign was started to create a new railway museum. Eventually, a building was provided in the form of the locomotive roundhouse at York North and in 1975 the National Railway Museum took over the former BR collection. Here are several locomotives positioned around one of the two turntables that were there when the museum was originally opened. (Author's collection)

❮ No 4472 *Flying Scotsman* is seen at Tyseley Depot, replenishing its water before continuing on to Derby for the next scheduled stop, on 27 May 2004. No 4472 *Flying Scotsman* and its support coach, No 17013, were working a positioning run from Southall to Doncaster, on the first leg of its transfer from Southall to the NRM at York. (Author)

^ No 4472 *Flying Scotsman* is at York ready to depart with a 'Scarborough Spa Express', shortly after becoming part of the NRM collection. (Adrian Scales)

❮ During June 2004 and July 2004, *Flying Scotsman* received various repairs prior to the 'Scarborough Spa Express' season. It is seen here awaiting one of the first trips of the year. (Adrian Scales)

❮ A classic railway scene with a classic locomotive, as No 4472 *Flying Scotsman* departs from Scarborough Station during 2004. However, a spate of failures with No 4472 necessitated including the NRM-owned former Royal Train locomotive No 47798 *Prince William* in the train's formation for insurance purposes. (Adrian Scales)

^ The boiler of *Flying Scotsman* rests on a well wagon at the works of Riley & Son during 2012, as remedial attention is given to the many cracks discovered on its frames. (Author)

^ Here the slide bars from *Flying Scotsman* receive rust-inhibiting paint at Riley & Son's during the spring of 2012. (Author's collection)

^ During its prolonged overhaul in NRM ownership, *Flying Scotsman* received many new components to get it back on the tracks again. A newly fabricated banjo dome is seen waiting to be fitted to the boiler. (Author)

› Riley & Son of Bury were contracted by the NRM to complete a vast amount of engineering work on *Flying Scotsman* to get it steaming again. Their name, together with the number 4472, had been stencilled on the boiler barrel after remedial work had been signed off as completed. (Author)

^ In 1973, Datsun (as it was called then, later becoming Nissan) launched its B-210 'Flying Scot' model, a version of the car better known in the UK as the Datsun Sunny. Whether potential buyers were really impressed that their new car could outrun *Flying Scotsman*, who can tell? (Author's collection)

› During 1972, the finances of *Flying Scotsman* were in such a poor state that the locomotive and its train were seized by the IRS in lieu of unpaid debts. One option considered by Alan Pegler was for *Flying Scotsman* and its train to relocate to the *Queen Mary* ocean liner complex, located at Long Beach, California, to be incorporated into a proposed restaurant area, but this did not come about. (Author)

‹ With works number No 1564, the first express passenger locomotive to be completed and make its debut for the newly formed L&NER left Doncaster Works, having cost £7,944 to construct. Here, that same locomotive – No 4472 *Flying Scotsman*, complete with Diamond Jubilee headboard and flags – is seen on the Carnforth turntable on 17 February 1983. (David Ward collection)

› During the non-stop 'Flying Scotsman' runs between King's Cross and Edinburgh Waverley, the footplate crews would change over through the corridor tender at Tollerton, north of York. As water is topped up, we see at close quarters part of the the end of the corridor tender attached to No 4472 *Flying Scotsman* during the McAlpine era. (David Ward collection)

‹ Numbered 60103 and in BR green livery, *Flying Scotsman* is seen working a demonstration 'Talisman' service on the Nene Valley Railway during the summer of 1994. (D. Trevor Rowe)

‹ At 05.20 on 18 May 1963, No 4472 *Flying Scotsman* departed from Gainsborough on its second public run since its acquisition by Alan Pegler, namely the Gainsborough Model Railway Society (GMRS) 1X50, the 'Isle of Wight Special'. *Flying Scotsman* is seen at Eastleigh Station waiting to return. (Author's collection)

^ No 4472 *Flying Scotsman* worked the GMRS' special on 12 September 1965, from Waterloo to Weymouth, via Polkesdown. It is seen being turned on the Weymouth turntable in preparation for its return working, which was via Yeovil into Paddington. (GMRS collection, Courtesy Joy and Richard Woods)

He had negotiated permission 'to run his engine continuously from 1963 to 1972, as a private engine hauling passenger trains', but despite the deal making no mention of preventing other engines doing the same on BR tracks, this agreement caused bad feeling with other engine owners who were not permitted the same freedom.

Despite being well within his rights to continue as he had been, Alan Pegler decided to take action before his contract with British Railways ended. He hit upon the idea of taking *Flying Scotsman* to the USA and Canada to work an exhibition train.

During the winter of 1968/9, No 4472 *Flying Scotsman* visited the Hunslet Engine Company's works in Leeds for a strip down and a heavy overhaul. This was to allow the railway authorities from the USA and Canada to inspect it to their satisfaction before it was authorised to run in North America. In further preparation for the trip, *Flying Scotsman* was again sent to Doncaster Works, where it was trial-fitted with a 'pilot' – otherwise known as a cowcatcher, an American-style buckeye coupling and a high-intensity headlamp. These were later fitted to the engine on board the *Saxonia* on its journey to Boston, but a bell, an American-style whistle and air brakes were fitted to 'Scotsman' at Doncaster, before it was painted in apple-green livery.

For the train which would accompany *Flying Scotsman*, there were five exhibition cars, a support coach and an observation car. Alan Pegler had agreed to deliver two British-built Pullman cars, *Lydia* and *Isle of Thanet*, to the National Railroad Museum in Green Bay, Wisconsin. These had been associated with General Eisenhower during the Second World War, as he toured various military facilities in Great Britain and on the European continent. The Pullmans would be used for accommodation while the train was touring North America. Also for 'delivery' were coaches 1591 and 1592, also associated with Eisenhower, originally built by the LNER in 1936 as first class sleeping coaches and which had their interior restored to a war-time appearance overseen by Bill McAlpine. These coaches, although shipped by Alan Pegler, didn't form part of the exhibition train and were forwarded to Green Bay, 'piggy-backed' on well wagons, once they were in the USA.

THE GAINSBOROUGH MODEL RAILWAY SOCIETY
and
ALAN PEGLER

THE FARNBOROUGH FLYER

SPECIAL BUFFET CAR EXCURSION
to

THE INTERNATIONAL AIR DISPLAY

Hauled by No. 4472

'FLYING SCOTSMAN'

Saturday, 10th September, 1966

From —
Doncaster, Mexborough and Conisborough - 70/-
Sheffield Midland and Chesterfield - - - 65/-
Derby Midland - - - - - - - 60/-

Juveniles 50/- any station

ROUTE VIA TRENT, LEICESTER, LUTON, BRENT, FELTHAM and WOKING to FARNBOROUGH. — 15 minutes walk to Air Display.
First Steam Loco South of Trent for several months.

LOCAL BOOKINGS
A1 SCOOTERS, 13, BALBY ROAD, DONCASTER (Tel. Don. 3846).
W. WEST, BRINKBURN LOW ROAD, CONNISBOROUGH
N. A. COOK, 325, ECCLESHALL ROAD SOUTH, SHEFFIELD (Tel. Sheff. 364272).
G. LEWIS, 14, NORTH ROAD, CLOWNE, CHESTERFIELD.

POSTAL BOOKINGS AND ENQUIRIES
S.A.E. to M. A. CLAPHAM, 69, BECKETT AVENUE, GAINSBOROUGH, LINCS.

PHONE BOOKINGS
G. HINCHCLIFFE, STOW. (Gainsborough) 311

Culdicotts Ltd., Gainsborough.

› On 10 September 1966, No 4472 *Flying Scotsman*, recently overhauled at Doncaster Works, worked a train to Farnborough and back. The trip was advertised as the 'first steam loco south of the Trent for several months'. (GMRS collection, Courtesy Joy and Richard Woods)

› On 21 May 1967, No 4472 *Flying Scotsman* worked a GMRS special, the 'Retford Rover'. The trip was organised by Alan Pegler along with the GMRS and cost 55*s*. (Author's collection)

So with all this in place, *Flying Scotsman* then added another achievement to its roll of honour, when during 1969, it crossed the Atlantic Ocean to work a 'British Trade Exhibition Tour Train' around the USA and Canada. Although the tour was initially successful, it eventually ran into financial difficulties and ended up on the frontage at Fisherman's Wharf in San Francisco as a static exhibit, with Alan Pegler being declared bankrupt in 1972.

On 13 August 1972, No 4472 *Flying Scotsman* and its train arrived at Sharpe Army Depot, Lathrop, California, for secure storage. Sharpe was very important during the Vietnam War and hundreds of army aircraft, both fixed-wing and helicopters, arrived at Sharpe to be prepared for shipment overseas. Twenty-four-hour operations began and Sharpe became the major depot for supplies moving westward to Southeast Asia. Unsurprisingly, the Sharpe Depot had considerable security in force, which probably explains why there are so few photographs of *Flying Scotsman* at this location.

This situation couldn't be allowed to last, however, so on 8 January 1973 a bill of sale for *Flying Scotsman* was drawn up, in which the locomotive, its two tenders and associated spares were to be purchased by Bill McAlpine for $72,500. A ship was found and No 4472 was moved to a Naval Supply Depot in Oakland.

○ ○ ○

Many readers with an interest in steam locomotives may well have heard about Alan Pegler, but what is known about him?

Alan Francis Pegler was born on 16 April 1920 and, after attending Cambridge University, he served in the Second World War as a Navy flyer and later as a member of the Royal Air Force, where he was a bomber pilot. In 1946, he joined the family business, the Northern Rubber Company, manufacturer of rubber and plastics, eventually becoming chairman and managing director. When the business merged in 1961 with other Pegler interests, Alan left the firm to devote his entire time to running the Welsh narrow-gauge Ffestiniog Railway, which he had bought in 1954 and of which he was chairman and subsequently president. He accepted an invitation to become a member of the Eastern Area Board of British Railways, on which he served from 1955 until 1963.

Beginning in the January of 1961, Alan Pegler spent eighteen months in negotiations with BR to buy *Flying Scotsman*, eventually persuading the owners to agree to the condition that he could run it continuously, from 1963 to 1972, on British Railways-owned lines. After successfully touring the country with *Flying Scotsman* for six years, he then took *Flying*

› This fine view of both of *Flying Scotsman*'s tenders was taken at Norwich during one of No 4472's visits there during the Alan Pegler era in the 1960s. (David Chappell)

^ A classic shot of Alan Pegler, the man who saved *Flying Scotsman* from destruction in the 1960s. On his driver's grease-top hat are the initials F and R, plus a coat of arms, signifying the Ffestiniog Railway – which Alan also saved from oblivion. (Penny Pegler collection)

Scotsman to North America on a trade mission, where he toured the USA and Canada and ended up at Fisherman's Wharf in San Francisco.

In 1971, Alan Pegler set up 'Flying Scotsman Enterprises' as a corporation based in Maryland, a move which made sense as he envisioned having *Flying Scotsman* out of the UK for some time. He had originally planned that *Flying Scotsman* would remain in San Francisco until the spring of 1972. About this time, he had received offers to exhibit *Flying Scotsman* in Japan, Australia and other countries.

Alan Pegler celebrated his ninetieth birthday on 16 April 2010 at the Ffestiniog Railway, with his friends and family around him. I was fortunate enough to be included in this gathering, in celebration of a man who had the vision to save *Flying Scotsman* for the nation.

Back in October 2008, I interviewed Alan about his recollections of *Flying Scotsman* at his home in Stepney, London and so I will let him tell his own story about No 4472 *Flying Scotsman*:

> The first time that I actually set eyes on 'Scotsman' was at the Wembley Exhibition and I do remember being taken with the green colour. Not very far away was one of the GWR locomotives – *Caerphilly Castle* – and at the time I thought that 'Scotsman' was far more attractive. It's all a matter of taste, actually.
>
> Then it got to the point of entering regular service and I do remember it was on display at Marylebone. I was genuinely interested in it, I don't know if it was the name that fascinated me, but it just became a fact of life that if anything special happened on the LNER then Gresley became involved. The name Gresley became very well known to me, I never met him but knew the name extremely well. When the first corridor tender appeared, the press releases showed 4472 from the rear end, showing the connecting vestibule door and so I was constantly reminded of 4472.
>
> It was numbered 1472 early on and then became 4472 and this was the number that I was looking for when I went out train-spotting – if I saw it, then that would be a bonus. Anyway, it finally came to the point that *Flying Scotsman* was not on the list of official locomotives for preservation. It wasn't just me, but a hell of a lot of railwayman of repute all thought that there was a great risk of scrapping *Flying Scotsman*. A great many people like me would have liked to have done something

^ One of the first times that the author became aware of *Flying Scotsman* was on 17 September 1966, when he saw a magnificently polished locomotive painted in stunning apple green, fitted with gleaming red nameplates working the last steam-hauled non-stop run from Victoria to Brighton. *Flying Scotsman* is seen passing through South London on that very day. (Author's collection)

^ *Flying Scotsman*'s first 'trade mission' in North America started in Boston on 8 October and ended in Houston, Texas, six weeks later. 'London Stateside Ltd' was founded in 1968 to export quality luxury merchandise to the USA. Ten specially selected girls, seen here in Battersea Park on 9 September 1969, accompanied the train acting as sales assistants on buses and as roving ambassadors for Britain in the United States. (Author's collection)

about it, but didn't know what to do. So, I got an original idea, no one put me up to it and so I thought if I buy the blinking thing and if it's a private engine then they won't be able to do what they want with it.

And so, I got to know a chap called T.C.B. Miller, known to everyone as Terry. He was on the BR Board and was British Railways Chief Engineer for Traction and Rolling Stock and he led the design team at Derby Railway Technical Centre, which produced the HST 125 train. He was very helpful to me and if there were 'rumblings from above', he smoothed them over. He took very kindly to me getting *Flying Scotsman* preserved. I believe I am correct in saying that he was an apprentice under Gresley.

So, a deal was done and I travelled on the footplate of 'Scotsman' on the 13.15 King's Cross to Leeds train on 14 January 1963, which was its last revenue earning service for British Railways. When it was uncoupled at Doncaster and went into the works, I then owned it.

The engine was in the works for only three weeks. Because Terry Miller and I had both decided that it would not be practical to put her back into original condition *à la* 1922–23, we decided between us that as near as possible we wanted to have her as a typical A1 class 'Pacific' in the mid-1930s, era and so a single chimney was fitted, even though I firmly believed in double chimneys.

^ No 4472 *Flying Scotsman* is ready for another turn of duty as it rests in its shed at Doncaster. It is fitted with a bell and whistle in preparation for its North American tour. (Frank Hornby)

^ No 4472 *Flying Scotsman* was lifted up from the shores of England ever so gently and ever so slowly on 18 September 1968 and was loaded on the starboard side of the *Saxonia*. After No 4472 had been secured to the deck, the pilot (cowcatcher), which had been made at Doncaster Works, was bolted to the front buffer beam, as was the buckeye coupling. Many feared that *Flying Scotsman* would never return home again. (Sir William McAlpine collection)

There was a story put around by the head of the paint shop that Gresley liked the idea that locomotives of distinction should have a red background to their nameplates. I got the feeling and still have, in my bones, that 'Scotsman' was his favourite locomotive and of course by that time, it was a locomotive of distinction, having been involved in different publicity events, and so I agreed to the red background.

With Terry Miller's help, I managed to negotiate a couple of trial runs with eleven coaches, from Doncaster to New England, Peterborough. 'Scotsman' didn't perform that well – but the idea was to stick to an un-streamlined 'Pacific' of the 1930s. Terry said that 'Scotsman' would 'need a proper overhaul sometime' and the time he recommended was 'not too far ahead' and that 'the overhaul would be at Darlington Works', well that was his suggestion anyway. The point of Darlington was that it wasn't going to be Darlington for much longer and if I wanted a proper major overhaul doing, then Darlington Works was the place to go in the not too distant future. So I went along with that and it was fine. In the course of telling me all about this he told me that there would be spares available, spare cylinders, spare boiler and other things – which I went along with and paid extra for the overhaul. So that is why we went to Darlington, Terry Miller had advised me that it wouldn't be a works for much longer and that there were people there who knew about Gresley 'Pacifics' and so it would be best if done at Darlington Works while it still existed – so that was how it worked out.

We had been running for some time and Terry Miller said that it was becoming more of a problem to provide water. I asked if it would be possible to acquire a second tender and convert it for water. Terry said yes, and that the extra water would take up the space where the coal went. It was all carried out with safety in mind and we ended up with two tenders and two corridors. The corridor tender was acquired from Scotland and this was also thanks to Terry Miller, but it was in poor condition and needed a lot of work doing to it and this work was also carried out at Doncaster.

There then came a very exciting time and for a period of twelve months, 'Scotsman' was the only locomotive allowed out on BR tracks. A lot of people became very jealous about it and by that time I had been elected to the Eastern Area Board and was with some pretty clued up businessmen. They made it clear that if I attempted to continue to run the locomotive, I would have to have a water tight agreement. In the agreement, I got permission to run the locomotive and anytime it was out on the main line, it was deemed, in the law as they interpreted it, as if it was still in the ownership of BR – that was how it worked.

Eventually, it became apparent that it was due for overhaul and duly entered Darlington Works. I kept well out of the way – they had been given a job to do by Terry Miller, I let them get on with it, he didn't encourage me to go there. The decision to paint the cylinder covers 'lined green' was carried out during the overhaul and it was a Darlington decision, I played no part in that whatsoever. But I liked it so much that I decided to leave the Darlington trademark when I went to America – that's how she went. I don't think people realised it was a Darlington trademark, but I thought it looked damn good, it was much more interesting than plain black.

After 'Scotsman' had been overhauled it went on display at an open day at Darlington Works, after which I did a trip for anybody who had been connected with its overhaul. An awful lot of people had stretched the point – anyone who had screwed something on 'Scotsman' came on the trip, but that was fine. The trip was from Darlington, down the main line as far as Doncaster, to Black Carr Junction, known as the 'Joint Line' down to Lincoln and March, where a phoney 'reservoir' was established to pick up water. Then on to Peterborough, where we missed our path and through to Doncaster and back to Darlington. The stock used was pretty much the 'Tees-Tyne Pullman' and the men on board had the 'full works'. My only other connection with Darlington Works was when *The Great Marquess* locomotive was overhauled there – I do remember that we ran double headed on one of the trips. Although not being very mechanically minded, I must say that Darlington Works did a superb overhaul – all extremely good, there were no problems at all, I was thoroughly satisfied with their workmanship.

Now the main thing that persuaded me to take 'Scotsman' to America was that the Great Western had gone before and the LMS, of course, had taken the *Coronation* and the *Royal Scot*. Other railways had done it – and that was the point, the LNER never had and I thought I would jolly well put the LNER on the map! It was as simple as that really. I think I am correct in saying that *Flying Scotsman* was the only LNER locomotive that went over. But, I wanted to take a train as well, in those days we had vacuum brakes and nobody else had, so the idea of taking a complete train was that everything would work and fit properly. For the train to be fitted out it went to Twickenham where there was spare siding space at that time. Just before the trip happened, 'Scotsman' went to Hunslet Works, Leeds. It was what the Americans insisted on, they said that they wanted it dismantled to ensure that it was in 'tip-top nick' – that the workmanship was good. So it was stripped down and put back together again and it was in excess of anything that was needed, nothing was replaced. The original whistle, which was fitted before the chime whistle in America, was what the Americans referred to as a 'peanut whistle' and the average American thought that it was pretty unexciting – which of course it was, but very typical of the LNER. The pilot, cowcatcher, was manufactured at Doncaster, it wasn't of conventional form and the Americans were expecting a more American type, but it did the job. As it was out of gauge on British Railway tracks, it was fitted on its journey to America. The bell, headlamp and whistle were donated by W. Graham Clayton who was a big fan of ours and was president of the Southern Railway system. He arranged to fly them over to us for fitting, the headlamp in particular is enormous and must have cost a fortune to air freight it!

It took three shipments to take the locomotive, two tenders and nine coaches to Boston, Massachusetts. The train consisted of five former LNER baggage cars – the exhibition coaches, two Pullman cars, one Pullman observation car and an 'administration car', a BR Mk I Brake Composite Corridor (BCK) coach, which carried the locomotive crew when they were not actually involved with the locomotive. The idea was that I would deliver the two Pullman cars to a railway museum in Green Bay. But the exhibitors in the exhibition cars needed somewhere to travel from A to B, for there was no passenger accommodation at all on the train. Pullman car *Isle of Thanet* stayed at the museum and the second Pullman car *Lydia* went right on to San Francisco.

When the train was finally dispersed, *Lydia* went back to the museum at Green Bay on a flatbed railway trailer. The baggage cars and the administration car were converted to become restaurant cars for use in Universal Studios and the observation car went to Fisherman's Wharf in San Francisco. I don't know what happened to the baggage and admin cars after their time at Universal, but I do know that the two Pullman cars and the observation car have come back to this country. With regards to crewing the locomotive and train on the first trip – well, Henry Foster, who was a King's Cross driver, who I had come across some way or other, impressed me very much in the way he handled the locomotive over here, was the driver and firing to him was David Court from Doncaster and we had two admin girls looking after the main body of 'Dolly Birds'.

Also, there was a chap who wanted to get in on the act and sell souvenirs of *Flying Scotsman* in America, but I was persuaded not to rely on the souvenirs to keep the engine running. I managed to get a couple of London double-decker-type buses and the idea was that they would run on the same route ahead of the train. But, we found that when it came to the crunch, the buses were very unsuitable – you couldn't get the people through a bus really, for proper sales they didn't work out very well and we decided to call it a day quite early on in New York. So, on the way down from Boston, there was the prospect of ditching a dozen or so girls more or less. I thought it would be bad publicity for the train and bad luck for the girls. So I said that we would take those girls on as well and this added a lot to the cost of the operation. I suppose that they were more trouble than they were worth, but they were very glamorous and added a lot of colour to the whole situation – very good!

We then made our way down to Houston in 1969, at the time that they were doing the 'Moon-Shots', and got involved with some of the people going to the Moon, with several of them coming along with us. They had not seen a main-line steam locomotive in their lives before and were very intrigued. We were in Houston for a week and had let the fire go out, so we had a ceremony to light up the fire again. One of the astronauts lit the match – which lit the oily rag to get the fire started, it all went very, very well. We then went from south of Texas to Green Bay, right through the middle of the United States, through unpopulated areas. On one occasion, one member of the crew said 'Stop the train here for a few minutes' and so we did exactly that and promptly blew the whistle and within a minute, people appeared from out of the ground!

And so we got up to Green Bay, Wisconsin and stayed there for nearly a month. They had great ideas there for running 'Scotsman' on their track around the museum area. Les Richards, locomotive inspector from York, said that '... there was no way that the locomotive was going around the track at the museum for fear of it getting derailed', much to his disappointment!

The leading tender was the one that carried the coal – about 9 tons – and one thing that staggered the Americans a bit was that we could go all the way from the Chicago area to Minneapolis, St Paul – 405 miles – and not need to pick up any coal on the way.

The old 'Scotsman' went very, very well and we were limited officially to 60mph. I gave a commitment that the observation car wouldn't exceed 60mph, but in actual fact I got it up to 75mph on the Canadian main line! But on the whole, it was adhered to.

After the first trip was completed I came back to England after the train had been put away at Slaton. I then returned for the second trip in 1970 and when this ended, the train was put away at Toronto until mid-summer, more or less. For the next trip the idea was to get to the West Coast to an established tourist area, make lots of money and it would be all right – the plan was to get to San Francisco and busk it!

Operations manager of the train George Hinchcliffe was promised things, but when it came to the crunch they just didn't happen! I managed

On 28 October 1969, No 4472 *Flying Scotsman* and its exhibition train departed from Washington for its next destination. The Washington Monument can be seen on the left. (Author's collection)

On 4 November 1969, No 4472 *Flying Scotsman* made a stop in the heart of Denison, Texas, and posed outside President Dwight D. Eisenhower's birthplace. Alan Pegler is seen leaning on one of *Flying Scotsman*'s buffers. (Paul Dowie collection)

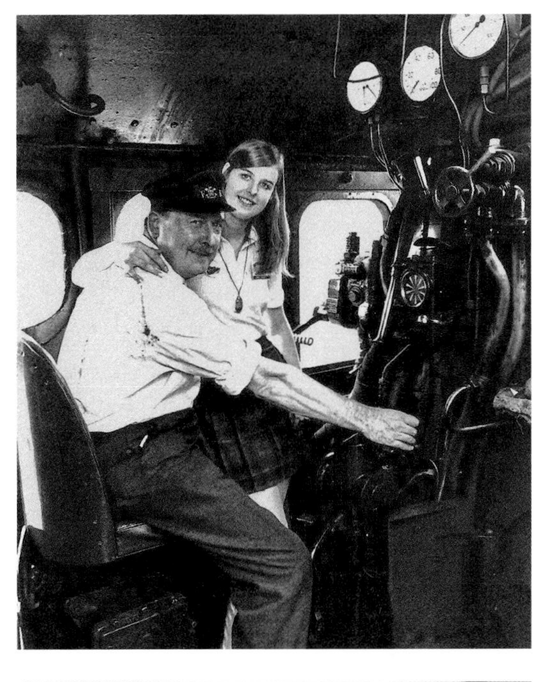

› Penny Pegler, daughter of Alan Pegler, is seen aged 17 with her father on the footplate of *Flying Scotsman* during a stop at Montreal, Canada, on part two of the North American tour. (Penny Pegler collection)

‹ A close-up view of the bell and whistle that were fitted to *Flying Scotsman* for its North American tour. The bell, which is owned by the author, bears the inscription: 'Donated by the Southern Railway System 1969'. (Author's collection)

› The Californian sunshine illuminates the driver's side of *Flying Scotsman*'s cab as it waits to traverse the Embarcadero in the early 1970s. The LNER coat of arms was hand painted in the gloom of Doncaster Works in March 1963. (Edward Saalig)

to negotiate space on Fisherman's Wharf – you could get seven coaches there. Some of the local shopkeepers said that it was a distraction to their local businesses and would be grateful when the train finally moved on. We did get a lot of visitors at that time, George got the impression that if we had got permission to stay on our original site on Fisherman's Wharf, we would have been able to break even enough to keep our heads above water. I decided that the only thing to do was to apply for bankruptcy – which I duly did and people started to take it seriously.

I was wondering what would happen to the engine and all the rest of it and luckily that's where good old Bill McAlpine came into the act. He liaised with George Hinchcliffe and took advantage of the time difference between east and west and he managed to get the deal done. So eventually it was just the case of getting the locomotive back to England. The exhibition coaches were just baggage cars and were of no historical importance at all, but I did manage to do a deal with a restaurant in Los Angeles, where they ended up more or less as restaurant cars, a most unlikely end for them, but I don't know if they survived.

My bottom line – it was, the whole thing, a fantastic experience and I wouldn't have missed it for the world! It's difficult to say I wouldn't do it again knowing what I know, but I met some fascinating people and saw lots of America that I wouldn't have seen otherwise. We came through a thing called 'Feather River Canyon', which is a very spectacular piece of railway – lots of things like that. I suppose that it altered my life in a large way, I put everything on the line to make sure that it did all happen. But eventually it all resulted in the engine being put up for sale publicly.

Thanks to Andrew Scott, they raised money to make sure that it didn't go to a foreign railway museum. It is having a major overhaul and will end up on trips to Scarborough, 40 miles on a flat attractive route and trading on the name. So for that it has worked out well.

When I got the OBE, the presentation was done by Prince Charles. The award was for getting the Ffestiniog Railway going and for keeping *Flying Scotsman* continuing to run. And so he said to me that 'In saving *Flying Scotsman*, was it true that I had bought the locomotive?' And I said yes, it was the only way to make sure that it would get preserved. He said 'You are a genius!' I have said many times to friends since then, that it was the one and only time that I would be called a genius by a Royal. I have been much tickled by that quite off the cuff remark. I haven't exactly dined out on that, but it has amused people very much. From my own personal point of view, I am here in East London, not in a nice country house in North Nottinghamshire. But, I've been awfully lucky with my family, many families would have said '... silly old devil, what did he blow all the family money on that for?' This whole scenario has kept what I regard as a part of British heritage in England and I'm not shedding any tears at all.

^ No 4472 *Flying Scotsman* is seen arriving in San Francisco on board the Western Pacific Railroads ferry *Las Plumas*, after it had docked at the San Francisco ferry slip, which was adjacent to Pier 41. The 'spike' above the number 72 on the buffer beam is the Transamerica Pyramid building. (Henry W. Brueckman, collection of Jack Neville)

• • •

On 18 March 2012, Alan Pegler, the man who saved *Flying Scotsman* from the breaker's yard for future generations to admire and enjoy, died aged 91 at his home in Stepney, East London.

7

THE SIR WILLIAM MCALPINE YEARS

'Mr Bill', as he is affectionately known to his railway friends, otherwise known as Sir William McAlpine Bt (Baronet), was the second saviour of *Flying Scotsman*. He was born in London at the family-owned Dorchester Hotel, and lived at the family home in Surrey until the age of 4. After his education at Charterhouse School he joined the family building firm at the Hayes Depot, Middlesex, which was busy during the post-Second World War national rebuild. Bill inherited the baronetcy in 1990 on the death of Lord McAlpine, 5th Baronet, and he is the patron of the Clan McAlpine Society. His name has also been immortalised on the EWS Type 60 diesel-electric locomotive No 60008, which he named *Sir William McAlpine*.

A well-known steam enthusiast, Bill returned to the Hayes Depot during the years of the Beeching axe, and found that the McAlpine Company's Hudswell Clarke 0-6-0ST No 31 had been sidelined for scrap, at the asking price of £100. So, Bill said, 'send it over to Fawley'.

With appropriate arrangements made, the locomotive, which marked its 100th anniversary in 2013, was moved to his country estate home at Fawley Hill, Buckinghamshire, where there wasn't so much as a railway sleeper in position – literally a 'greenfield site' ready for development. As anyone who has been fortunate enough to visit will know, Fawley Hill lives up to its name in terms of gradient, and the only way in was across a neighbour's field. The low-loader got stuck in the mud and as it was urgently needed elsewhere, the saddle tank was unloaded on to a track panel, one of the two Bill had acquired. Over the next two weeks, the tank engine was edged forward by pulling the track panels one in front of the other with a bulldozer. Nothing is ever simple in running a railway!

This saw the beginning, in 1961, of the Fawley Hill Railway, a private railway which now runs to over a mile in length and which boasts the steepest gradient at 1:13 on a British railway. It features a magnificent collection of railway items, which include:

- Passenger and goods railway rolling stock, including a Royal coach
- A magnificent railway museum
- The GER station that originally stood at Somersham in Huntingdonshire
- A MR signal box from Shobnall Maltings – the maltings were originally built by the world famous Bass Brewery at Burton upon Trent
- The footbridge from Brading on the Isle of Wight – as Bridge No 25, it spanned the Ryde Pier to Shanklin railway line

The railway line itself is embellished with many prominent architectural and decorative features which Bill has gathered over the years. These include early cast-iron bridge parapets, the original Wembley Stadium concrete

Having safely returned to British shores, and waved on by thousands of people who lined the trackside, *Flying Scotsman* ran under its own power from Edge Hill to Derby Works. After *Flying Scotsman* had been inspected, it was intended to repaint it in its former BR dark-green livery, with the former number 60103 being restored, but, in fact, it was painted in apple green once more. (Sir William McAlpine collection)

An unusual view of the nameplate on the driver's side of *Flying Scotsman*, taken at Carnforth during the 1980s. (Sir William McAlpine collection)

flag poles, and arched structures which originate from well-known London railway locations. The most recent addition is an 'iron-henge' formed of cast-iron columns that were made redundant when St Pancras Station was recently developed. Being a private railway, entrance to Fawley Hill is by invitation only and these are usually given during summer months.

Bill McAlpine had been a member of the Transport Trust, which was given the task of finding a replacement for the Clapham Transport Museum. After searching diligently they found a suitable site at Crystal Palace, only for ministers in the Edward Heath government to decide that the new museum had to be outside London, with the result that it moved to York. Prince Philip was invited to open both the North Road Museum at Darlington, built by Bill's firm, and the new National Railway Museum on the same day, 27 September 1975. Bill's problem was that he had to be at North Road when HRH left and at the NRM before he arrived on the Royal Train. Bill tried to cadge a lift on the Royal Train but 'security' disallowed it, so he called in the firm's helicopter, which on a misty day faced a bit of trouble getting airborne, but in the end he made it just in time – only to discover that the train bringing the other dignitaries from London was 90 minutes late!

Around this time Bill considered buying the railway line in the Dart Valley area of south Devon with the intention of lifting the track and laying a narrow gauge railway line. He hired a BR diesel multiple unit (DMU) for £100 and went to scout out the area. He came to the conclusion that the railway would be best left as standard gauge, as indeed it remains to this day.

Bill acquired various coaches in his career, in particular when he and Pete Waterman bought out the BR Special Trains Unit – quite a train set amounting to 200 coaches! His true love in railway coaches is probably with GE No 1, the directors' saloon for many years; Bill was able to attach this to main-line trains being hauled by No 4472 all around the country.

After starting the Fawley Hill Railway, Bill acquired GWR 'Castle' class 4-6-0 locomotive No 4079 *Pendennis Castle*, with John, later Lord, Gretton of Bass, Ratcliff & Gretton brewery fame. Bill owned No 4472 and No 4079 at the same time, reuniting the two locomotives that had been displayed together all those years ago at Wembley. He originally kept them at Market Overton, where it was hoped to establish a steam heritage centre. Unfortunately BR made life difficult by realigning the East Coast Main Line and main-line access became impossible.

After the efforts of Dr Peter Beet to preserve the LMS motive power depot at Carnforth, formerly designated 10A, Bill became a shareholder there from 1974 and then had a controlling interest in it until 1987. So after the failure of the Market Overton scheme, Bill transferred both

› 'Man and Machine'. Sir William McAlpine is seen posing with his locomotive *Flying Scotsman* at Marylebone Station, just before its departure with yet another enthusiasts' special during the 1980s. (Sir William McAlpine collection)

No 4472 and No 4079 to Carnforth, where Bill established Steamtown, the railway visitor attraction.

Interestingly, before their closure the workshops at Market Overton performed the sectioning work on 'Merchant Navy' class 'Pacific' No 35029 *Ellerman Lines*, which is now in the NRM, York. Various parts that were missing were cannibalised from another Bulleid 'Pacific' in NRM ownership, namely No 34051 *Winston Churchill*.

Built for the GWR loading gauge, No 4079 was rather too wide for some of the BR network and was limited to working on only a few routes around the country, so a decision was made and No 4079 was sold to the Rio Tinto Company who moved it to Australia.

Bill also became involved in a plan to save the Romney, Hythe & Dymchurch Railway, where he became its chairman. He had led the rescue bid in 1972 by getting twenty of his friends each to contribute £5,000 to buy enough share capital to get the railway back on its feet again, with the resulting success that is seen today.

So, how did 'Mr Bill' come to be the proud owner of *Flying Scotsman*?

In early 1973, following the bankruptcy of Alan Pegler, the speculation was that unless something was done then No 4472 could be lost to America permanently, or at worst even be cut up. Alan Bloom, owner of a renowned garden centre in Norfolk, made a phone call to Bill to discuss the matter.

This led Bill to ask George Hinchcliffe, who had run *Flying Scotsman* for Alan Pegler in North America, to fly to San Francisco to see what the situation was. George said that the attorney handling the sale had died, and that if Bill wanted to buy *Flying Scotsman* then a deal could be done, but that he would have to act fast to meet American deadlines. George added that he could make arrangements to return the locomotive

7 September 1989

Guiness Publishing Company
33 London Road
Enfield
Middlesex

Dear Sir,

LOCOMOTIVE NO.4472 EX LNER "A3" PACIFIC
"FLYING SCOTSMAN"
WORLD RECORD NON-STOP RUN

This is to certify that the Steam Locomotive No.4472 Flying Scotsman completed a NON-STOP run between Parkes and Broken Hill in New South Wales, Australia on 6 August 1989.

* Total distance covered without stopping = 422 miles 7.59 chains.
* Total elapsed time whilst in motion = 9 hrs 25 mins 15 secs.
* Total load hauled by the Locomotive = 535 tonnes.

We, the undersigned, travelled on this train and verify that the above facts are true and correct.

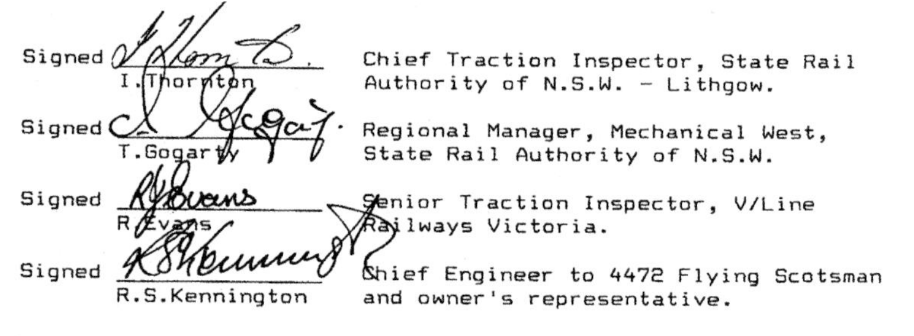

Signed I.Thornton — Chief Traction Inspector, State Rail Authority of N.S.W. - Lithgow.

Signed T.Gogarty — Regional Manager, Mechanical West, State Rail Authority of N.S.W.

Signed R.Evans — Senior Traction Inspector, V/Line Railways Victoria.

Signed R.S.Kennington — Chief Engineer to 4472 Flying Scotsman and owner's representative.

home, providing finances were in place. Bill then asked George that if the finance was arranged and *Flying Scotsman* was returned home, would he be willing to run the locomotive in Britain? George said that he would. This was a critical decision, for if George had been unable to arrange for *Flying Scotsman* to return home, then it would have faced an uncertain and precarious future in North America – with the very real possibility of it going for scrap.

But the good news was that No 4472 did return to the UK in a journey which took it by ship via the Panama Canal in February 1973. It remained in such good condition that the BR authorities said that *Flying Scotsman* would be permitted to run with its two tenders and under its own steam from Liverpool to Derby Works. At Derby, the paintwork scorched by the Californian sun was removed and No 4472's paintwork was restored to its former glory, becoming the first LNER apple-green livery that Derby had ever applied.

After its overhaul at Derby, No 4472 moved to the Paignton & Dartmouth Steam Railway for the summer of 1973, after which it was transferred to Market Overton in Lincolnshire, where it was reacquainted with its rival from Wembley, No 4079 *Pendennis Castle*. After a spell there, *Flying Scotsman* was transferred to Steamtown at Carnforth, from where it worked various tours around the country. 1986 saw the next major change, as Bill McAlpine decided to lease a two-road former diesel maintenance shop at Southall, which became the new base for No 4472.

During the autumn of 1987, a suggestion came from the NRM to Bill McAlpine, saying that as they were unable to release world speed record holder *Mallard* for shipment to Australia for the country's bicentennial

^ During *Flying Scotsman*'s Australian tour, on 8 August 1989, it made a non-stop run from Parkes to Broken Hill. To everyone's enormous excitement, *Flying Scotsman* created a new world record (as seen in this authentication document) for a non-stop run by a steam locomotive of 422 miles 7.59 chains, in 9 hours 25 minutes, with 535 tons gross. Seven British and Australian drivers were used, as were three 10,000-gallon water gins. (David Ward collection)

› The words on the plate say it all. (Roland Kennington)

celebrations twelve months hence, perhaps Bill would be interested in sending No 4472 *Flying Scotsman* instead …

Walter Stutchberry and his Aus Steam '88 colleagues burned the midnight oil in discussion, raised the finance and finally agreed the complete package for the visit with Bill's people in England. With the paperwork completed, everything was done to ensure that *Flying Scotsman* was ready for its trip 'down under'. The work included: refitting tyres on to the coupled wheels; the fitting of air brake equipment to ensure tenable running in Australia; repainting of the locomotive; and finally, undertaking the test runs on British Rail to ensure that everything was in working order before *Flying Scotsman* left these shores in September 1988.

When asked how he felt about *Flying Scotsman* going to Australia, bearing in mind what happened when 'Scotsman' went to North America, 'Mr Bill' would have been within his rights to be apprehensive. However, he said that he was very excited indeed at the thought of taking *Flying Scotsman* to the other side of the world, as they had learned from previous lessons and had in place the money to bring her home before she left the country!

In October 1988, *Flying Scotsman* arrived in Australia and during the course of the next year travelled more than 28,000 miles, including a transcontinental run from Sydney to Perth. One of the central attractions of Aus Steam '88 involved *Flying Scotsman* double-heading with New South Wales Government Railways locomotive No 3801 and running parallel alongside two Victorian Railways 'R' class locomotives along the 190-mile-long stretch of broad and standard gauge tracks of the North East railway line in Victoria.

No 4472 stayed in Victoria for two months before heading back to New South Wales. On 8 August 1989, *Flying Scotsman* set another world record by travelling 422 miles from Parkes to Broken Hill non-stop, the longest such run ever recorded by a steam locomotive.

During September 1989, *Flying Scotsman* travelled to Perth, at the very extremity of Western Australia, and on an occasion reminiscent of the days of the American Wild West, *Flying Scotsman* met 'smoke box to smoke box' with its old sparring partner *Pendennis Castle*! *Flying Scotsman*'s adventures in Australia turned out to be not just a visit but an extended tour lasting some fifteen months, until No 4472 arrived back in the Port of London, at Tilbury Docks in December 1989. Once again No 4472 *Flying Scotsman* had returned safely to the UK.

No 4472 *Flying Scotsman* was then overhauled at the former steam depot of Southall (81C) between December 1989 and May 1990, during which the air brakes and electric lights were removed.

On 2 May 1990, complete with the commemorative headboard that it had carried on its world record non-stop run to Alice Springs, No 4472 *Flying Scotsman* made its first main-line run on BR tracks following its return from Australia. The route was from London Paddington to Banbury, where it stopped for water and then continued via Acocks Green to Carnforth. *Flying Scotsman* then worked a series of specials around the country, including some 'Cumbrian Coast Express' charter specials.

From April–July 1993, No 4472 *Flying Scotsman* was at Babcock Robey's works for a heavy repair and before the overhaul was completed, Sir William McAlpine said that he was happy for *Flying Scotsman* to be painted in BR dark-green livery, as a thank you to the volunteers who were looking after it. During its time there, Babcock's managing director would visit the shop floor every day to see how the overhaul was progressing. On one occasion, he spoke to Roland Kennington, the chief engineer of *Flying Scotsman*, and said that when the locomotive's overhaul was completed, he wanted something really spectacular to finish off the overhaul and to create much publicity. Roland had for some time wanted to retrofit a double chimney and German-style smoke deflectors. So, when the managing director heard about this, he was absolutely delighted and offered to put up the money for these and a new smoke box to be fitted. So the two ideas came together and *Flying Scotsman* had the new smoke box, double chimney and smoke deflectors fitted, and was duly out-shopped in BR dark-green livery – more correctly known as 'middle chrome green' – with its running number changed to No 60103, the number it carried at the end of its BR career. In this form, *Flying Scotsman* received much acclaim among enthusiasts and the public alike.

In its new guise, on 25 July 1993, No 60103 *Flying Scotsman* recommenced its tour of preserved railways, starting with its second visit to the Paignton & Dartmouth Railway, before moving in September 1993 to visit the Gloucester & Warwickshire Railway. As the touring continued, Sir William McAlpine decided he would have to sell *Flying Scotsman* to pay off the mortgage on the locomotive. Pete Waterman approached Sir William and offered to merge his railway interests with those of Sir William's. As a result, on 21 September 1993, 'Flying Scotsman Enterprises' and 'Waterman Railways' merged to form 'Flying Scotsman Railways'. In the new company Pete Waterman and Sir William McAlpine each had a 50 per cent share of No 60103. With Sir William and Pete each involved in the ownership of the new company, between them it was agreed that Pete would run the business side of the new enterprise.

8

THE PETE WATERMAN YEARS

At this time, *Flying Scotsman*'s BR certificate had expired and, as it had recently been overhauled, beginning in April 1994 the engine spent the next few years running on preserved railways, commencing a stay at the Nene Valley Railway. Disguised as fellow classmate No 60106 *Flying Fox*, it was used on demonstration freight trains for photographers.

No sooner had the ink dried on the contract with Sir William that Pete started having some doubts about whether his involvement with *Flying Scotsman* was such a good idea. He started receiving hate mail over what was seen as 'his decision' to repaint *Flying Scotsman* in BR green livery instead of its former LNER apple-green livery. He even received a threat from a member of the clergy, which stated that unless Pete repainted *Flying Scotsman* quickly into apple green, then he would personally tie Pete on to the railway tracks and drive *Flying Scotsman* over him as a punishment. The truth of the matter, of course, was that the decision to paint 'Scotsman' had been taken long before Pete had become involved with it.

So far we have seen that Pete Waterman is renowned in the railway arena, and to this day he remains closely involved in the efforts to reopen the Gloucestershire & Warwickshire Railway, but he is, of course, more widely famous for the earlier stages of his career. Dr Pete Waterman OBE DL is known the world over for his success in music production, and when I visited him at his offices and recording studios located in the old County Hall building in Central London, I was overwhelmed by the number of gold, silver and platinum discs displayed on the walls. Over 200 hit singles were represented, from the repertoire of such stars as Kylie Minogue, Rick Astley, Bananarama and Steps.

Music aside, Pete Waterman's main interest undoubtedly lies in railways and he has been a part of several railway projects over the years. For instance, in 1988 he revived the name of the London & North Western Railway Company for his rail vehicle maintenance business. This Crewe-based company was the largest such business in the country and was sold to Arriva in November 2008. Currently, Pete owns a steam locomotive and carriage restoration company called LNWR Heritage, which is based at Crewe Heritage Centre.

Pete also founded the Waterman Railway Heritage Trust, looking after several steam, diesel and electric locomotives, including the GWR locomotives 2-8-0T No 5224 and 2-6-2T No 5553, the last steam engine to leave Woodham Brothers scrapyard in Barry, South Wales in January 1990, which Pete has stated is his favourite locomotive. Other steam locomotives have included 4-6-0 No 7027 *Thornbury Castle* and BR

^ Painted in its former BR green livery and with its BR running number of 60103, *Flying Scotsman* continued to work on preserved railways in this form between 1993 and 1996. It is seen working 'The Talisman' service on the Nene Valley Railway. (D. Trevor Rowe)

❮ Pete Waterman, on the left, was a 50 per cent owner of No 60103 *Flying Scotsman* with Sir William McAlpine during the 1990s. He is in the cab with Roland Kennington, chief engineer, during No 60103's visit to the Swanage Railway during September 1994. (Andrew P.M. Wright)

Standard 4-6-2 No 70000 *Britannia*. His diesel locomotives have included Type 46 No 46035 *Ixion* and Type 47 No 47402 *Gateshead*. Pete has a fondness for the original 25kV AC electric locomotives, which he regularly travelled behind for many years – they are certainly the type to 'spark' an interest! He was approached by BR at Derby and bought 81002, 82008, E3035 and 85101. Among the many other locomotives over the years that Pete's Heritage Trust has saved for posterity is Beyer Garratt locomotive No 109, originally from South Africa. Constructed in 1939 at their works in Gorton, Manchester, this locomotive was the first of the class to be built by Beyer Peacock and was the first NG/G16 locomotive to be repatriated to the UK. With the restoration carried out at the LNWR workshops in Crewe, the restored locomotive will run on the Welsh Highland Railway in North Wales.

As well as the full-sized and fully working versions, Pete also has a special interest in model railways and founded 'Just Like the Real Thing' – a model railway business that specialises in O-gauge scale kits. To this day, he remains invested in the company, regularly accompanying its sales stands to model railway exhibitions. As could be imagined, Pete is the

^ That 'inner sanctum', that 'hallowed ground' where all train enthusiasts want to go – the cab of a steam locomotive. Of course, for many there is none finer than *Flying Scotsman*. (Author)

^ On 22 June 1995, the dismantling of *Flying Scotsman* was authorised in preparation for its next major overhaul. Here No 60103 is seen inside its home shed awaiting the work to commence. (Fred Stenle)

owner of an extensive collection of railway models and layouts both in O scale and larger gauges. As of 2013 he is building an O scale model of the 1950s Leamington Spa Station. In addition to actual modelling, he has also written extensively on the subject. Pete received an OBE for his services to music shortly after his 57th birthday.

On 14 July 1994, No 60103 *Flying Scotsman* commenced its stay at the Swanage Railway, working passenger services on the Dorset coast. A couple of months later, in September, it was transferred to the Severn Valley Railway and then moved on to the Birmingham Railway Museum. All was going well with *Flying Scotsman* on its 'grand tour' of private preserved railways around the country, and it later moved on to the Llangollen Railway in Wales.

On one occasion on this heritage line, No 60103 *Flying Scotsman* was being put away after its day's duty when, due to an error by the signalman, it suffered a derailment of all wheels at low speed. After being re-railed and checked it resumed its duties. However, on 28 April 1995, disaster struck again! No 60103 *Flying Scotsman* was dramatically and hastily withdrawn from service. Steam was seen escaping at about head height from the 'back head' of the boiler into the fireman's side of the cab. *Flying Scotsman* was deemed a total failure and was immediately withdrawn from service. The locomotive had run approximately 30,500 miles on nine preserved railways since October 1992.

On 6 June 1995, No 60103 *Flying Scotsman* started its journey by road from the Llangollen Railway to its home depot at Southall, West London. Arriving at the West London waste terminal in Brentford, it was then put on to the tracks of the truncated Brentford branch line and was moved by rail to its home depot at Southall. Dismantling was authorised in preparation for its next major overhaul, but the process was slow and lacked enthusiasm due to the spiralling outlay that would clearly be needed to overhaul *Flying Scotsman* to be able to return it to working order.

Fortunately, Dr Tony Marchington, an acknowledged steam fanatic, saw an article in a railway magazine that stated that *Flying Scotsman* had put up for sale at an asking price of £1.5 million. He approached Sir William to discuss the matter, so Sir William agreed that he would handle the sale. Much to the half-owner's surprise, Tony Marchington didn't haggle at all and agreed to buy *Flying Scotsman* at the full asking price. With a deal agreed, No 60103 *Flying Scotsman*'s next major overhaul was resumed in earnest.

9

THE TONY MARCHINGTON YEARS

Salvation for *Flying Scotsman* had arrived for a third time when Dr Tony Marchington bought the locomotive and had a three-year restoration undertaken to return it to main-line running condition. Tony was the man who owned the ultimate collection of 'big boys' toys', a magnificent range of traction engines and fairground equipment that suggested there was little more left to relish. The old question 'what do you buy for the man who has everything?' could easily have applied to Tony – especially once *Flying Scotsman* was added to his collection!

As well as setting up and owning Oxford Molecular, a company which provided chemical information management and decision support software to researchers in the pharmaceutical and other related chemical research industries, he headed the Buxworth Steam Group, which was a family organisation that offered old-time fairground attractions for corporate hire. There were twenty-five traction engines, including a showman's engine called *The Iron Maiden*, which starred in the eponymous 1962 film. In addition, there were many fairground rides such as: the 'Steam Galloping Horses'; a helter-skelter; a Ferris wheel; a cakewalk (the old Victorian favourite involving traversing a moving walkway); a 1925 set of German-built 'Chairoplanes'; and various other rides. There was a 'Wall of Death', complete with its 1930s Indian motorcycle, which used to be ridden by a larger-than-life character called 'Tornado Smith', who rode it complete with a sidecar in which he carried his pet lion.

Tony had always had a passion for steam locomotives, including railway locomotives, saying: 'I had always had the wish to own an A4 streamliner – magnificent machines.' Then one day, as Tony was waiting at Oxford Station, he noticed on the front of a railway magazine that *Flying Scotsman* was up for sale. 'I telephoned Bill McAlpine and we struck a deal,' he said. Tony then went on to buy the A4 class 'Pacific' No 60019 *Bittern* so that there was another locomotive available to run excursions when *Flying Scotsman* was being serviced – or so the theory went.

After a three-year rebuild, *Flying Scotsman* worked 'The Inaugural Scotsman' from King's Cross to York on 4 July 1999. With the world's media and thousands of well-wishers lining the track side to observe the successful event, Tony then authorised the running of excursion trips around the country.

After many such excursions, No 4472 *Flying Scotsman* was dragged 'dead' over the Brentford branch line, from its Southall Depot to Brentford waste terminal, West London on 25 August 1999, prior to low-loader transportation to Hartington Moor Traction Engine Rally for its first visit there. Upon its return to Southall Depot a week later, the foundation

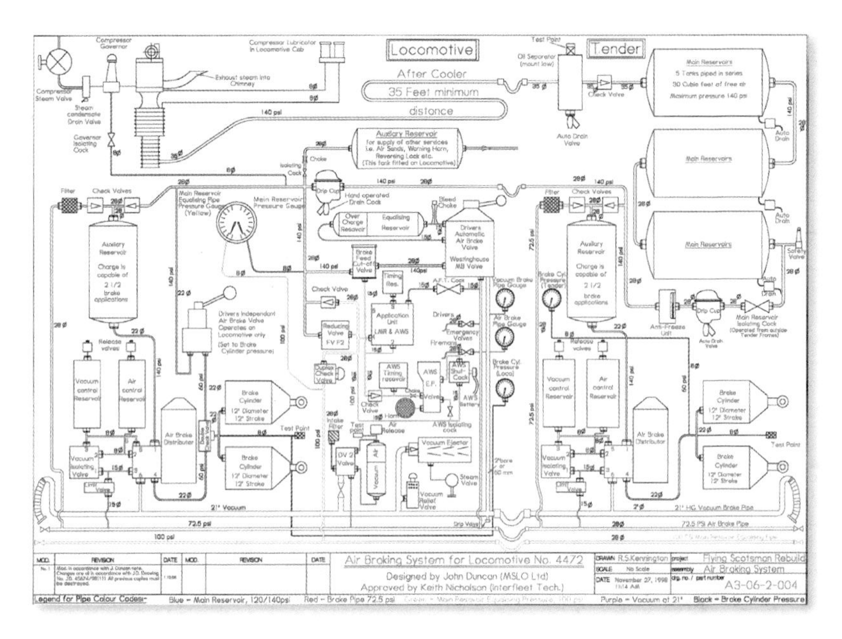

❮ And you thought that the mechanics of a steam locomotive were simple! Dated 1 December 1998, here is a revised version of a schematic diagram of the air braking system for No 4472 *Flying Scotsman*, as used under Tony Marchington's ownership. (Roland Kennington)

ring was washed out and *Flying Scotsman* continued its duties working excursions around the country.

On 6 February 2000, No 4472 *Flying Scotsman* was substituted for 'Merchant Navy' class 4-6-2 No 35028 *Clan Line*, which was unavailable, and departed from Platform 2 at London Victoria with its first run with the British Pullman section of the Venice Simplon-Orient-Express (VSOE) to Southampton. Indeed it worked many VSOE excursions all over the country. But the one exception was the Kent routes, for here *Flying Scotsman* was banned from working, as it was deemed to be 'out of gauge' working under certain bridges and structures on some routes.

On 17 October 2000, No 4472 *Flying Scotsman* then departed from St Pancras Station at 10.08 with day one (of a five-day charter) of the 1Z98 'Royal Scotsman', destined for Inverness. The 'Royal Scotsman' train provided luxury accommodation for customers paying £3,300 for a one-way journey, taking in York, Edinburgh and Perth en route to Inverness. It was organised by Holland & Holland, the sport and travel agency, using the 'Royal Scotsman' luxury train.

Flying Scotsman continued working excursions and on 4 April 2001, No 4472 *Flying Scotsman* departed from Platform 2 at Victoria with the first train under a new five-year contract to work trains of the British section of the Venice Simplon Orient Express, working VSOE's 'British Pullman' and 'Northern Belle' luxury trains within the UK. VSOE officials described the trains as iconic and representing all that was best in British engineering, which naturally made *Flying Scotsman* a perfect fit.

In December 2001, it was announced that a visitor centre dedicated to the story of the world's most famous steam locomotive was set to be built in Edinburgh during the next couple of years.

In 2002, as *Flying Scotsman* saw regular use on the VSOE British Pullman trains, investors seeking their own bit of steam train history were given a little more time to 'get on board', after managers of *Flying Scotsman* announced it was extending the deadline for applications for shares in the famous train's £2 million flotation. It was also stated that money raised from the flotation would fund a new multi-million-pound theme park in Doncaster. Dr Marchington also maintained that he wanted to open a *Flying Scotsman* visitor attraction on a site next to Waverley Station in Edinburgh, which could become its semi-permanent home.

In the event, more than 1,000 train enthusiasts bought shares in *Flying Scotsman*, raising £850,000. The amount of cash raised meant that the company then had an extra £1.95 million at its disposal and on the completion of the share issue, £1.1 million of debt was also converted into equity.

Flying Scotsman plc joined the OFEX exchange for smaller firms on 13 March 2002, with the money, or so it was planned, to turn the locomotive into a profitable business. With the company valued at around £4.7 million, Peter Butler, the chief executive, said: 'Today marks the beginning of a new chapter for the company and enables us to capitalise on the huge potential of the Flying Scotsman.'

But the signs were not promising. On 25 September 2002, an article in the *Edinburgh Evening News* stated that, in its first set of interim results since floating, Flying Scotsman plc had posted a pre-tax loss of £276,000 for the six months to the end of June, compared with a £320,000 deficit for the same period the previous year. A total of £60,000 was spent on repairs and maintenance, slightly lower than the previous year's £61,000. Turnover rose to £122,000 from £73,000 and sales were set to rise further in the second half of year.

^ No 4472 *Flying Scotsman* prepares to depart from Platform 2, Victoria Station, with a VSOE British Pullman train, during 2002. (Frank Hornby)

Unfortunately, in 2003, Edinburgh City Council turned down the plans for a visitor attraction, to be known as 'Flying Scotsman Village', and in September of the same year Tony Marchington was declared bankrupt and consequently was required to sell off his prized collection of traction engines and fairground rides. At the company's AGM in October 2003, CEO Peter Butler announced losses of £474,619, and, with a £1.5 million overdraft at Barclays Bank, stated that the company only had enough cash to trade until April 2004. After Flying Scotsman plc failed to declare interim results, the company's shares were suspended from OFEX on 3 November 2003.

In February 2004, it was announced that a vintage car dealer had been drafted in to seek potential buyers for the historic locomotive. The debt-ridden company admitted that it might sell No 4472, mainly due to the lack of progress with the proposed 'steam village' scheme. The car dealer, based in Oxfordshire, had sent details of the locomotive to some of his customers to gauge potential interest, although with a guide price of £2.5 million believed to have been suggested, interest may well have been low. However, Geoff Courtney, spokesman for Flying Scotsman plc, said that the chance of the locomotive being sold abroad was 'absolutely minimal', because of its importance to Britain. Geoff refused to comment on claims that the sale had been prompted by fears that a major bank might seize and auction the locomotive to recoup debts.

'*Flying Scotsman* had last run in January 2004 and was undergoing its winter overhaul before trips were to resume in March,' said Peter Butler, the company's chief executive, a former Milton Keynes MP, Oxford solicitor and Oxfordshire County Councillor. Coincidentally, he was a classic car enthusiast himself. He went on to say that 'the accounts show that No 4472 *Flying Scotsman* is valued at more than £2 million. We have had some interest but I can't say yet whether they are from people with money. The value is far higher than just the locomotive.'

In the event, Flying Scotsman plc asked international property advisors GVA Grimley to hold a sealed bidding process, effectively auctioning the legendary locomotive.

During this period, various rescue bids to save *Flying Scotsman* started to appear, including one by the National Railway Museum entitled 'SOS – Save our Scotsman'. On 1 April 2004 NRM bosses put in their bid to keep the world's most famous steam locomotive running on Britain's railways. They had managed to submit their offer a day before the 2 April deadline after a campaign to raise enough money to buy *Flying Scotsman*. The bid included more than £425,000 raised through a public appeal, as well as an undisclosed but 'substantial' donation from Sir Richard Branson on behalf of the Virgin Group. Andrew Scott, head of the NRM, said:

> The level of support and the strength of feeling towards this historic locomotive has been absolutely staggering. There can be no doubt that, without the generous donations and goodwill of the British public, we could not even have hoped to put together a worthwhile bid. Our staff have handled more than 5,500 calls, letters and emails since the appeal was launched and we would like to thank everyone who has contributed to the fund. If our bid is successful, we hope to invest any donations received after today in ensuring that the locomotive is fit for main-line operations.

The outcome of the sale was announced on 5 April, when it was revealed that the NRM had been successful in its bid to buy *Flying Scotsman*. Because of the general public's active interest in saving *Flying Scotsman*, it became known as 'the People's Engine'.

Very sadly, on 16 October 2011, Dr Tony Marchington died of cancer, aged 56. With standing room only, some 250 friends and family attended his funeral, which took place at St James' church, Buxworth, on 28 October 2011. Tony had bought *Flying Scotsman* at a cost of £1.5 million and, after a three-year restoration costing him a further £1 million, No 4472 *Flying Scotsman* returned to working on the main line in 1999. When it was sold to the NRM in 2004, it was certainly in better condition than when he had bought it in 1996 – a truly special legacy to the nation.

10

THE NATIONAL RAILWAY MUSEUM YEARS

With *Flying Scotsman* put up for sale in 2003, the high-profile national campaign headed by NRM chief Andrew Scott led to the classic locomotive being bought in April 2004 by the NRM. In the process it became part of the National Collection at York.

The NRM's public appeal raised an initial £365,000, which Sir Richard Branson's Virgin company matched. A further £60,000 from public donations raised the total to £790,000, leaving enough funds to keep the locomotive running in the UK for years to come. The successful bid was put forward with the help of a £1.8 million grant from the National Heritage Memorial Fund, a government-backed body with the remit of preserving buildings and other objects of importance to Britain's national heritage. The fund's chairman, Stephen Johnson, speaking shortly after the purchase on 5 April 2004, said: 'We are absolutely chuffed … there will be a lot of chuffed people today, including everyone who cares about Britain's heritage and all at the NRM, and York's tourist bosses.' The primary aim of the purchase and funding agreements was to return *Flying Scotsman* to main-line operation. No 4472 *Flying Scotsman* then made its debut in National Railway Museum ownership between 29 May and 6 June 2004, appearing at the 'Railfest' event at the museum in York.

Flying Scotsman then ran for some twelve months, with interim running repairs, to raise funds for its ten-yearly major boiler recertification. In 2005, No 4472 *Flying Scotsman* was scheduled to work one of the NRM's York to Scarborough steam excursions, but due to a problem with one of its superheater elements it was withdrawn from service, with work put in hand to repair it as soon as possible. Duly repaired, it returned to service and worked various specials around the country, including 'Christmas Lunch' specials from Dorridge to Leicester.

Then in January 2006 an NRM press release stated that No 4472 *Flying Scotsman* had entered the NRM's workshops for her major main-line overhaul. Within a short time, *Flying Scotsman* was rapidly dismantled to such an extent that the running plate was the only component recognisable to the casual observer. In connection with the overhaul of *Flying Scotsman*, it was decided to revert to using the A3 class pattern boiler, which was most recently the locomotive's spare. The A4-type boiler, used on the 'Pacific' during its last period in steam, was sold to raise funds for *Flying Scotsman*'s overhaul. Work included the turning of the driving wheel tyres at Tyseley Locomotive Works in Birmingham, while workshop staff at the NRM replaced the bottom half of the cab sides.

Over the ensuing years, the remedial work continued and then, on 21 June 2009, the boiler barrel of No 4472 *Flying Scotsman* was noted at the workshops of Riley & Son at Buckley Wells, Bury, resting on a well wagon. Devoid of its smoke box and firebox, it nevertheless still retained

❮ The Museum of British Transport opened in 1963, located in the disused Clapham Bus Garage, and featured old buses, trams and fire engines, as well as locomotives. Among those on display was the Furness Railway's 1846-built 0-4-0 locomotive No 3 *Coppernob*. (Author's collection)

❯ When the Museum of British Transport closed in 1972, the railway exhibits were relocated to Leeman Road in York and the NRM was opened in 1975. It was an immediate success with the public, and continues to be so today. (Author)

the banjo dome. The front tube plate and parts of the throat plates were in the workshop waiting to be fitted, with the firebox in the process of having stays fitted. On the same day, the corridor tender and wheeled frames of No 4472 were in the workshop at the NRM in York.

After almost fifteen years of service, Andrew Scott announced on 26 June 2009 that he would be retiring as director of the NRM in late 2009. Andrew's time at the NRM would be remembered for him leading the NRM team that successfully bought *Flying Scotsman* for the nation, making the NRM the most visited museum in Britain outside of London, and for winning many awards, including 'European Museum of the Year'. The recruitment process for finding Andrew's successor quickly came up with the replacement – Steve Davies MBE, a strong leader with a passion for railways. The announcement was made on 1 October 2009 and he took up his position in February 2010.

Later that year, work on the boiler of No 4472 *Flying Scotsman* was progressing well at Bury, while the frames were away from the NRM workshops at York, having the air braking system completed.

The choice of livery worn by *Flying Scotsman*, as we have seen, is an emotive subject among some of those involved in the preservation of historic rolling stock. *Flying Scotsman*, and more particularly its owners, have attracted more than their fair share of criticism as a result of over fifty years in preservation. Alan Pegler's preferred option was to return *Flying Scotsman*, as far as was possible, to the general appearance and distinctive colour that it carried at the height of its fame in the 1930s.

A decision was later taken in the 1990s to reinstate the Kylchap double blast pipe, double chimney and German-style smoke deflectors that it carried at the end of its career with BR in the 1960s. All of these encourage more complete combustion, a factor in dealing with smoke pollution and fires caused by spark throwing. In this form it was painted in its former BR dark-green livery and was renumbered as 60103.

During the Tony Marchington era, although still fitted with a double chimney, it reverted to apple-green livery and, later still, it had the smoke deflectors refitted. This resulted in a great deal of adverse comments, as the colour and deflectors were inconsistent. It never ran in this form during its LNER days in apple green.

With the NRM's overhaul of *Flying Scotsman* finally coming to an end during the summer of 2011 – or so it was thought – it was decided that, while *Flying Scotsman* was completing its test runs, it would be painted in Second World War 'war-time' black livery and not the customary undercoat grey during this period. A former number carried by *Flying Scotsman*, 103, was painted on the driver's cab side and another former number, 502, was painted on the fireman's cab side, with the war-time austerity letters 'NE' painted on the tender. *Flying Scotsman* was again fitted with a double chimney and German-style smoke deflectors, certainly an anachronism when compared with the black livery. However, it was in this guise that the locomotive was unveiled to a specially selected group of dignitaries on 27 May 2011. The plan was that, after the testing programme had been completed, the more favoured livery of LNER apple green would be reapplied, but it would still be fitted with a double chimney and German-style smoke deflectors, and again this caused lots of negative opinion among rail enthusiasts. Those that wanted the authentic dark green for *Flying Scotsman* as fitted with double chimney and smoke

❮ The boiler 'back-head' from *Flying Scotsman* at the workshops of Riley & Son during the spring of 2012. The 'back-head' is the external wall of the firebox which protrudes into the cab of the locomotive and is where the operating controls are mounted. (Author)

⌄ The number 103 is seen painted on the cab of *Flying Scotsman* at the works of Riley & Son in Bury. *Flying Scotsman* originally carried the number 103 and 'war-time' black livery between May 1946 and January 1947. (Author)

deflectors would have to wait a while longer before the locomotive was in authentic livery once more.

After the public launch, *Flying Scotsman* was sent to Bury to have finishing touches applied; however, new cracks were found in the frames and the locomotive had to be stripped down once more for remedial work. This raised doubts that the protracted overhaul, started in 2006, would ever be finished satisfactorily. In truth, this setback was simply the latest in a long line of problems that the restoration project had faced. These included: problems with the original carried boiler; the need to manufacture additional parts for the replacement boiler; cracks in the horn blocks and frame stretchers; misalignment of the frames; and the need to manufacture new horn ties, to name but a few.

By the end of October 2012, the work being undertaken by the NRM engineering team included the fitting of a newly manufactured bogie stretcher, the overhauling of bogie components, the manufacture and fitting of ash-pan components, the manufacture and fitting of the cab floor and the overhaul and fitting of the lubrication system. The new GSM-R (Global System for Mobile Communications – Railway) cab radio system had also recently been commissioned and installed. The relentless remedial work continued, with no end date in sight for *Flying Scotsman* to steam again in the foreseeable future.

On 21 September 2012, it was announced that the NRM's director Steve Davies was to step down after just two and a half years in the role, to pursue a new venture in the private sector. He left his post at the end of October 2012.

Steve presided over some very high-profile projects at the NRM, not limited to the ongoing restoration of *Flying Scotsman*. He oversaw the launch of a purpose-built art gallery and arranged 'Railfest 2012', which included the biggest ever gathering of 'rail record holders'. He was also the driving force behind arranging the return of two exiled A4 class

❯ An artist's impression of *Flying Scotsman*, as it would appear for its relaunch in May 2011. The number 103 is on the fireman's side of the cab and number 502 is on the other. As can be seen from the previous picture, the number 103 is on the opposite side of the cab to that on which it was originally proposed. No 4472 *Flying Scotsman* was renumbered as No 502 on 20 January 1946, and on 5 May 1946 it become No 103. (Justyna Snigurska)

locomotives, No 60008 *Dwight D. Eisenhower* and No 60010 *Dominion of Canada*, from the USA and Canada respectively, for the historic and probably never to be repeated line-up of all six surviving A4s, planned for July 2013 to mark the seventy-fifth anniversary of No 4468 *Mallard*'s world record run on Stoke Bank in 1938. Steve also commissioned a report that was to be written in response to the prolonged delays and rising costs of restoring *Flying Scotsman,* which, although bought by the museum in April 2004, still hadn't turned a wheel in anger since its restoration had started in earnest in January 2006.

Steve's position was taken by Paul Kirkman, who arrived at the NRM on 5 November 2012 at the beginning of a year-long secondment from the Department for Culture, Media and Sport.

On 26 October 2012, the NRM's report on *Flying Scotsman* was published. It had been written by Bob Meanley, chief engineer at Vintage Trains, Birmingham, assisted by Professor Roger Kemp of Lancaster University, a fellow of the Royal Academy of Engineering. The report began with a damning indictment of the spiralling costs: '*Flying Scotsman* began a major overhaul in January 2006 that was scheduled to last one year and cost around £750,000. In October 2012, the cost of the overhaul had risen to around £2.7 million and hadn't been completed.'

The inquiry reviewed all aspects of the restoration with the aim of applying recommendations to the final stages of the project and future projects of this nature. The report revealed that '... the purchase of *Flying Scotsman* by the NRM would always have gone ahead, regardless of the locomotive's actual condition, given the aspiration to save it on behalf of the nation'.

The NRM acquired the locomotive in a sealed-bid auction, where there was no opportunity to negotiate on price. The report concluded that the museum's bid price of £2.3 million had been reasonable. However, the report continued that:

> ... even though *Flying Scotsman* had run intermittently until December 2005, the condition of the locomotive on purchase was poor ... it had had a large number of owners, several of which had failed financially. It had been heavily used and maintenance standards had been neglected. The restoration of this iconic and historic locomotive is one of the most complex engineering projects of its type ever to be undertaken ... The conservation desire to retain as much of the original locomotive as possible and to use British workmanship has also presented challenges, not least because parts have to be individually made.

The report, however, drew the following conclusions:

> ... Several issues added to the delays and costs. These include the absence of a detailed investigation either when it was purchased in April 2004 or soon after. This would have highlighted that it was in a much worse state of repair than was believed and identified the serious structural defects that were only recently found. As a result, a restoration project that was always going to have been complex and taken many years was given an unrealistically short time frame and budget at the outset.
>
> ... Other major challenges have been faced relating to the project management and engineering expertise, [including] the fact that the heritage railway engineering sector is 'a cottage industry' and disruptions caused by staff changes and illness. There have also been conflicts between the need to balance the requirements of the refurbishment programme with the museum's commitment to enabling the locomotive to be seen and enjoyed by the public.

The report made a number of recommendations regarding how the museum should go about the final stages of the project and any future projects of this nature.

Three senior managers, who were Tony Marchington's right-hand men during Tony's ownership of *Flying Scotsman*, hit back at the scathing criticism of the condition that the former LNER 'Pacific' was said to be in when it was sold to the NRM in 2004. Peter Butler, who was Tony Marchington's chief executive during his ownership of No 4472, said, '... we never made any assertions as to the condition of the locomotive – it was sold as seen.' Roland Kennington, chief engineer to *Flying Scotsman* 1986–2004, said, '... between 1999 until 2004, when it was sold to the NRM, *Flying Scotsman* ran more than 28,000 miles all over the country. In that time it had only two failures, in July 2002 and July 2003 and in both cases there were no delays to other train services.' This last point was also highlighted by David Ward, operations director for the 'Pacific' throughout the Marchington Years. He said, '... two failures in 28,000 miles represents 14,000 miles per casualty, which is commendable.'

The report's ultimate conclusion was that '... until the working group and engineering consultants have had more time, the National Railway Museum is not making any further announcements about any return to steam date, although it can confirm it will not be this year [2012]'.

Whenever *Flying Scotsman*'s overhaul is completed, it is hoped by those at the NRM that it will once again be pulling in crowds of admirers for many years to come, ensuring that the many millions of pounds spent to get this icon of railway history moving again have not been wasted.

EPILOGUE

Because of the LNER's emphasis on using *Flying Scotsman* for publicity purposes and then its ever continuing antics in preservation, including its two forays on to international shores, *Flying Scotsman* is surely without doubt the most famous steam locomotive in the world. As such, it has been featured many times in fiction, on screen and in print.

No 4472 *Flying Scotsman* made a short appearance in an episode of the 1960s spy show *Danger Man* in the episode called 'The Sanctuary'.

Then, No 4472 *Flying Scotsman* was featured in 'The Railway Series' books written by the Revd W. Awdry, featuring *Thomas the Tank Engine* – in the minds of the nation's children, at least, probably another contender for the title of the most famous steam locomotive in the world! In the book *Enterprising Engines*, No 4472 *Flying Scotsman* visited the fictional Island of Sodor in 1967–68, to cheer up his only surviving brother *Gordon*. At this time *Flying Scotsman* had two tenders and this was a key feature of the plot of one of the stories, called 'Tenders for Henry', where *Henry* was jealous. When the story was filmed for the television series *Thomas & Friends*, only *Flying Scotsman*'s two tenders were seen poking out of a shed, but both of the tenders had a coal space, which is obviously incorrect as *Flying Scotsman*'s second tender only carried water. *Flying Scotsman* was to have played a larger role in the series, but there is a popular rumour that the model was damaged and the budget was not sufficient for the modelling crew to rebuild the whole engine.

^ The LNER emphasised *Flying Scotsman* by using it for publicity purposes. It made its film debut in *The Flying Scotsman*, with the necessary shots achieved using a camera mounted on a flat wagon being propelled by another steam locomotive. (Author's collection)

^ No 60103 *Flying Scotsman* in a torrential downpour. At this time it had a single chimney, left-hand drive and was paired with tender No 5640. This LNER-designed, eight-wheeled non-corridor streamlined tender ran with *Flying Scotsman* from 2 July 1938 until 15 January 1963. (Author's collection)

On 11 January 1969, No 4472 *Flying Scotsman* and its exhibition train were on display at Terminal Station, Atlanta, Georgia, where the builder's plate and whistle are seen here in close-up. (The Southern Museum of Civil War and Locomotive History, Kennesaw, Georgia collection)

In 2000, No 4472 *Flying Scotsman* was featured in the Disney film *102 Dalmatians*, where it is seen pulling the 'Orient Express' out of London. The film stared Alice Evans, Glenn Close, Joan Gruffudd, Gérard Depardieu and Tim McInnerny. No 4472 *Flying Scotsman* was coupled to a selection of Pullman cars from the VSOE's 'Orient Express' train, normally based in London, and an electric generator car – which actually spent more time on screen than either the Pullman train or *Flying Scotsman*. A week was spent plying between St Pancras Station and Stewarts Lane Depot, which lies in the shadow of the former Battersea Power Station, to allow filming to take place throughout the night. As the filming happened around midnight, powerful Pinewood Studio floodlights transformed the station to give the impression of daylight. The usual organised chaos of a film being made ensued, including steam generators and model 'stand-ins' for the puppy Dalmatians being provided on the platform's edge. In the final cut of the film, *Flying Scotsman* appears for just 5 seconds – such extravagance in the film industry is probably quite normal! After the week's filming had finished, No 4472 *Flying Scotsman*'s final departure proved to be the last ever departure of a steam locomotive from St Pancras Station before it was closed for redevelopment as the major terminal for international train services through the Channel Tunnel. This was one of the less heralded records set by *Flying Scotsman*.

On 14 February 1973, mantled by snow after a rough transatlantic crossing, No 4472 *Flying Scotsman* was unloaded from the deck of the container vessel *California Star* by the floating crane *Mammoth*. The return journey to Liverpool had been via the Panama Canal. (Sir William McAlpine collection)

Flying Scotsman poses on the turntable at Carnforth on 23 April 1977. (T. Boustead)

› No 4472 *Flying Scotsman* is seen at the end of Platform 1, King's Cross Station, ready to depart with another excursion shortly after its overhaul in Tony Marchington's ownership. (Author's collection)

‹ Here are three of *Flying Scotsman*'s owners at an event at the NRM. On the left is Alan Pegler OBE FRSA, owner 1963–72; next to him is Tony Marchington, owner 1996–2004; and on the right is the Hon. Sir William McAlpine, Bt, FRSE FCILT, owner 1973–96. (Sir William McAlpine collection)

At the beginning of the new millennium, No 4472 *Flying Scotsman* was featured in a PC game called 'Microsoft Train Simulator'. To achieve a realistic effect of the locomotive, software developers from the Microsoft company travelled over from California to *Flying Scotsman*'s base in Southall, West London, to conduct a measuring and photographic survey of the locomotive. They also travelled on some of the locomotive's trips to record authentic 'puffing' sounds that were incorporated into this popular game, which is still available.

On 14 May 2003, *Flying Scotsman* was due to work a non-stop charter for Flying Scotsman Railways, in association with the VSOE, from King's Cross to Edinburgh Waverley, to mark the seventy-fifth anniversary of the first non-stop 'Flying Scotsman' service on 1 May 1928. The trip had been organised by Roland Kennington, chief engineer of *Flying Scotsman*. Roland had persuaded Venice-Simplon Orient Express to provide a rake of 'Northern Belle' coaches for the run. All engineering work along the route was to be suspended to allow the train access for a non-stop run along the main line. The logistics of the trip took quite a lot of planning: in order to avoid having to stop, the concentrated weight of three tenders carrying a total of 20,000 gallons of water needed to be hauled, meaning the weight capacity of all bridges on the route had to be checked. The two additional tenders required for the trip were to have been the second tender for *Flying Scotsman*, which was being completed at Southall, and the tender

› 'A Rake Of Railwaymen' – each of the people in this photograph played a major role in the running of *Flying Scotsman*. Taken on 17 May 1969 in Bressingham Gardens, we see (from the right): Alan Bloom, smoking his pipe, who later helped Bill McAlpine to repatriate *Flying Scotsman* from California; David Ward, driving the locomotive, operations director for *Flying Scotsman* 1996–2004; Alan Pegler, the saviour of *Flying Scotsman* in 1963, alongside his wife (facing the opposite direction); and finally, on the extreme left is George Hinchcliffe alongside his son Richard. Under Alan Pegler's ownership, George was No 4472's tour manager in North America and negotiated a deal with outstanding creditors after the period at San Francisco. He also arranged for *Flying Scotsman* to be shipped home, with Bill McAlpine financing the whole operation. They are all enjoying a ride on No 25, a former Beckton Gas Works 0-4-0 locomotive, at Bressingham Gardens Railway. (David Ward collection)

belonging to A4 class locomotive No 60009 *Bittern*. Unfortunately, even with all these plans in place, the trip was cancelled because of a misunderstanding with the availability of the passenger carriages required, but what a magnificent trip this would have been – even in the modern era *Flying Scotsman* still engenders interest in planning daring escapades such as this.

No 4472 *Flying Scotsman* was the first choice for BBC Television's *Top Gear* in its special 'Race to the North' episode, which involved a locomotive, a motorcycle and a car racing each other from London to Edinburgh. However, as No 4472 was dismantled and unavailable due to its overhaul taking place at the NRM, the starring role was given to a new-build steam locomotive, No 60163 *Tornado*, which was constructed using the Peppercorn A1 class design. Ultimately, the Jaguar car with James May driving won the race, though *Tornado* set a number of steam records in the process of coming a close second.

Flying Scotsman was available in its A3 class form, in its LNER-era apple-green livery, as an OO-gauge working model. This scale model

^ A stylised image of *Flying Scotsman*, produced for a major postcard producer during the early 2000s. (Derek Crunkhorn)

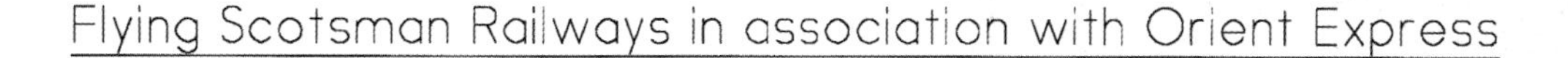

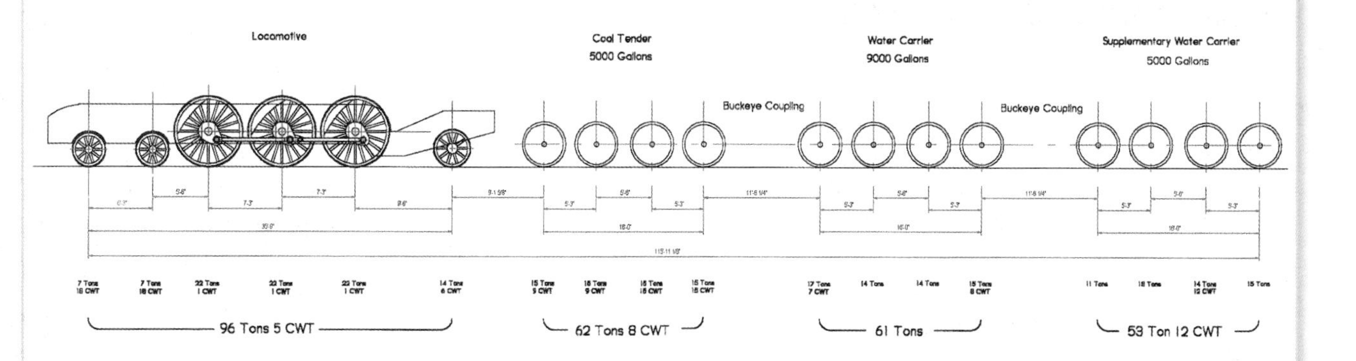

› 14 May 2003 was set aside for *Flying Scotsman* to work a non-stop run from King's Cross to Edinburgh Waverley, marking the seventy-fifth anniversary of the commencement of the regular non-stop 'Flying Scotsman' service. Here is the proposed arrangement using three tenders, providing adequate water for the run. (Roland Kennington)

was part of a special limited edition of 1,000 pieces produced by Hornby, exclusively for the NRM. The model featured a double chimney and came with smoke deflectors, a limited edition certificate and was DCC ready. DCC, or digital command control, is a control system for a model railway, whereby a decoder on board the locomotive is used to provide control of speed and direction of the motor; hence the railway modeller controls the locomotive and not the track.

○ ○ ○

Everyone with an eye for railway beauty and the poetry of motion has long admired *Flying Scotsman*, the superb creation of Sir Nigel Gresley. Almost from the time that the locomotive was first constructed, the LNER cared for No 4472 *Flying Scotsman* with reverence and pride, and from the time the paint was barely dry the locomotive attracted the loving attention of the public.

No 4472 *Flying Scotsman* became a symbol of power and prestige during the lifetime of the LNER. Unfortunately, after the nationalisation of British Railways in 1948, company loyalties were gradually dissipated and, with the demise of steam traction on BR, the eventual fate of the vast majority of Britain's steam locomotives became only too apparent. Perhaps surprisingly, when viewed from the modern boom era of heritage and museums, it was left to the embryonic railway preservation movement to ensure that *Flying Scotsman* survived, at a time when economics rather than sentiment dictated official preservation policy.

Fortunately Alan Pegler saved *Flying Scotsman* for the nation, and indeed the world. Subsequently, Bill McAlpine took over that responsibility in 1973 and, after Pete Waterman had a hand in running *Flying Scotsman* in the 1990s, that responsibility then passed on to Tony Marchington in 1996. The current owners at the National Railway Museum stepped into the breach in 2004, and at the time of writing continue to do their best to preserve the locomotive. In years to come, let us hope that *Flying Scotsman* is not known as 'the most famous steam locomotive in the world' for the expense and time taken to get it moving once more, but because of its exciting and heart-stopping achievements of the past.

Who knows what further exciting adventures await *Flying Scotsman* in the years to come?

^ Steve Davies MBE was director of the NRM from February 2010 until October 2012, and was responsible for authorising *Flying Scotsman* to be painted in 'war-time' black livery during its 'running-in' period after overhaul. Steve is with the frames of *Flying Scotsman* on 30 September 2011, at the workshops of Riley & Son in Bury. (Author)

› *Flying Scotsman* is seen in 'war-time' black livery at the NRM's 'Railfest 2012' event. (Author)

BIBLIOGRAPHY

BOOKS

Allen, Cecil J., *The Locomotive Exchanges: 1870–1948*, Ian Allan Publishing, 1950

Atkins, Geoffrey and Banks, John, *Railway Portraits: 50 Years of British Railway Development in Contemporary Photographs*, Venture Publications, 2002

Bagwell, Philip S., *Doncaster: Town of Train Makers 1853–1990*, Doncaster Books, 1991

Blake, Tom, *Prime Rib and Boxcars: Whatever Happened to Victoria Station?*, Tooters Publishing, California, 2006

Brown, F.A.S., *Nigel Gresley: Locomotive Engineer*, Ian Allan Publishing, 1961

Chacksfield, John, *Sir William McAlpine: A Tale of Locomotives, Carriages and Conservation*, The Oakwood Press, 2009

Clay, John F. (ed.), *Essays in Steam*, Ian Allan Publishing, 1982

Clifton, David, *The World's Most Famous Steam Locomotive: Flying Scotsman*, Finial Publishing, 1997

Dudley, John and Muter, Michael, *Flying Scotsman: On Tour – Australia*, Chapmans, 1990

Dunstone, Denis, *Welsh Railways: A Photographer's View*, Gomer Press, 2002

Fischer, Tim, *Trains Unlimited: In the 21st Century*, ABC Books Australia, 2011

Garratt, Colin, *The Complete Book of Locomotives*, Hermes House, 2004

Harris, Michael, *British Main Line Services: In the Age of Steam 1900–1968*, Oxford Publishing Co., 1996

———, *Gresley's Coaches: Coaches Built for GNR, ECJS and LNER, 1905–53*, David & Charles, 1973

Horton, Glyn, *Horton's Guide to Britain's Railways in Feature Films*, Silver Link Publishing, 2007

Hughes, Geoffrey, *Sir Nigel Gresley: The Engineer and His Family*, The Oakwood Press, 2001

Jones, Robin, *World's Fastest Steam Railway*, Morton's Media Group Ltd, 2012

Kichenside, G.M., *British Railways Coaches*, Ian Allan Publishing, 1962

Lambert, Anthony, *Lambert's Railway Miscellany*, Ebury Press, 2010

Larkin, Edgar J., *An Illustrated History of British Railways' Workshops*, Heathfield Railway Publications, 2007

McNicol, Steve, *Flying Scotsman: Profile*, Railmac Publications, 1992

Nock, O.S., *The British Steam Railway Locomotive: 1925–65*, Ian Allan Publishing, 1966

Parkin, Keith, *British Railways Mark I Coaches*, Atlantic Transport Publishers, 1991

Peel, Dave, *Locomotive Headboards: The Complete Story*, The History Press, 2010

Pegler, Alan, Allen, Cecil J. and Bailey, Trevor, *Flying Scotsman*, Ian Allan Publishing, 1969

Pike, Jim, *Locomotive Names: An Illustrated History*, The History Press, 2009

Siviter, Roger, *The Settle to Carlisle: A Tribute*, Bloomsbury Books, 1984

Townend, P.N., *East Coast Pacifics at Work*, Ian Allan Publishing, 1970
Whitehouse, P.B. (ed.), *The Last Parade – An Authorised Tribute to British Steam Preservation*, New Cavendish Books, 1977
Yeadon, W.B., *Yeadon's Register of LNER Locomotives, Volume One: Gresley's A1 and A3 Classes*, Irwell Press, 2001
———, *Yeadon's Register of LNER Locomotives, Appendix One: Named Engines, Their Application, Derivation and Changes*, Book Law Publications, 2003
———, *Yeadon's Register of LNER Locomotives, Appendix Two: Locomotives Tender Numbering*, Book Law Publications, 2005
———, *Yeadon's Register of LNER Locomotives, Volume Four: Gresley V2 and V4 Classes*, Book Law/Railbus Publications, 2001
———, *Yeadon's Register of LNER Locomotives, Volume Three: Raven, Thompson and Peppercorn Pacifics*, Book Law/ Railbus Publications, 2001

PERIODICAL PUBLICATIONS

Belper News
British Railways Illustrated
Doncaster Free Press
Edinburgh Evening News
Heritage Railway
Lubbock Avalanche Journal
Meccano Magazine
National Railway Museum Review
Oxford Mail
Railroading
Railway World
Steam Railway
The Lawyer
The Locomotive Magazine
The Locomotive Club of Great Britain monthly journal
The Press, York
The Railway Magazine
The Scotsman
The Scotsman on Sunday
Trains
Trains & Railways
Trains Illustrated
Yorkshire Post

WEBSITES

http://southern.railfan.net/
www.corusservices.com
www.sirnigelgresley.org.uk/old-index.html
www.sixbellsjunction.co.uk
www.vintagecarriagestrust.org/

MUSEUMS AND ASSOCIATIONS

The Museum of the Southern Railway Historical Association, USA
The National Railroad Museum, Green Bay, Wisconsin, USA
The National Railway Museum, York
The Southern Museum of Civil War and Locomotive History, Kennesaw, Georgia, USA